TOTAL ENVIRONMENTAL COMPLIANCE

A Practical Guide for Environmental Professionals

TOTAL ENVIRONMENTAL COMPLIANCE

A Practical Guide for Environmental Professionals

Bert P. Krages II

CRC Press
Taylor & Francis Group
Boca Raton London New York

CRC Press is an imprint of the
Taylor & Francis Group, an **informa** business

CRC Press
Taylor & Francis Group
6000 Broken Sound Parkway NW, Suite 300
Boca Raton, FL 33487-2742

First issued in paperback 2020

ISBN 13: 978-0-367-57902-9 (pbk)
ISBN 13: 978-1-56670-319-2 (hbk)

Visit the Taylor & Francis Web site at
http://www.taylorandfrancis.com

and the CRC Press Web site at
http://www.crcpress.com

Library of Congress Cataloging-in-Publication Data

Krages, Bert P.
Total environmental compliance : a practical guide for environmental professionals / by Bert P. Krages II.
p. cm.
Includes bibliographical references and index.
ISBN 1-56670-319-0 (alk. paper)
1. Environmental engineering—United States. 2. Environmental law—United States. I. Title.
TA170.K73 2000
628—dc21 99-089523
CIP

Library of Congress Card Number 99-089523

Preface

Prior to becoming an attorney, I actively practiced as an environmental engineer. This technical background continues to serve me well when assisting clients in matters involving compliance with environmental laws; however, I have noticed that many environmental professionals who seek advice on legal issues are confused about the legal principles underlying environmental requirements and how they apply to specific situations. The primary purpose of this book is to help environmental professionals understand how environmental laws function so they can become more effective with respect to matters that involve compliance concerns. Rather than discuss specific laws, the book focuses on what environmental professionals and their organizations can do to set a framework and implement practices intended to achieve total environmental compliance.

This book eschews the notion that substantial compliance is adequate, because, with few exceptions, environmental law does not recognize the concept of "substantial compliance." Furthermore, the history of environmental enforcement in the U.S. strongly suggests that many parties have suffered because their perception of what constituted adequate measures did not conform to the perceptions of the agencies regulating their operations. Rather than guess at what can be left undone, the book encourages organizations to strive to meet all environmental requirements and provides concrete guidance on how to meet this goal.

While this book is oriented toward providing environmental professionals with knowledge they can use to improve their skills in regulatory matters, the guidance it provides will be useful to anyone whose career involves working with environmental laws. Among the persons who may benefit from the book are business managers seeking to improve their understanding of compliance issues, consultants who specialize in activities affected by environmental requirements, and attorneys who advise clients on matters involving environmental laws. Environmental regulation and enforcement are becoming even more pervasive and intense. By developing their skills to the degree necessary to cope successfully with existing and emerging requirements, environmental professionals and others can reasonably expect to meet the challenge of achieving total environmental compliance.

About the Author

Bert P. Krages II is an environmental attorney who has advised clients on issues throughout the United States. Prior to entering law, Mr. Krages was a practicing environmental engineer. He received a B.S. in Environmental Engineering from Northwestern University, an M.S. in Environmental Engineering from the University of North Carolina at Chapel Hill, and a J.D. from the University of Oregon School of Law. He is admitted to the Oregon State Bar, U.S. District Court for the District of Oregon, and the U.S. Court of Appeals for the Ninth Circuit. In addition, he is registered as a professional engineer by the State of California.

Contents

Dedication

To my wife

KATHRYN PYLE KRAGES

1 Compliance, Noncompliance, and the Enforcement Process

Why is it important to achieve environmental compliance? The answer is simple. Organizations that consistently comply with environmental laws maintain control over their operations and finances. In addition, their employees will have more time to focus on activities other than environmental matters and need not fear the prospect of criminal liability. Achieving total environmental compliance will allow organizations and their employees to function more effectively and efficiently even though their activities are highly regulated.

The fact that compliance is a wise course of action does not mean that it is easy to achieve. Organizations that want to reap the benefits of compliance not only must want to comply with environmental laws but also must come to understand the factors that make achieving compliance difficult. Once these factors are known, organizations can set up a framework that helps them avoid the pitfalls that prevent others from achieving total environmental compliance. This framework, when combined with practices that promote communication, preparedness, and professionalism, will enable organizations to experience a synergism that makes achieving environmental compliance easier over time.

Although many people contend that it is impossible to comply with environmental laws all the time, it can be done if organizations are persistent and plan around the requirements that apply or will apply to their activities. Most people who advocate that compliance is impossible seem to believe that the sheer number of laws and permit requirements make it inevitable that at least some operations will miss compliance issues. Others contend that ambiguous or technically unsound laws make compliance impossible because regulated parties cannot accurately discern the requirements that apply to their activities. Although these arguments reflect the difficulty in attaining complete compliance, such problems can be overcome with foresight and active management of regulatory issues.

1.1 THE BENEFITS OF ENVIRONMENTAL COMPLIANCE

In theory, everyone should benefit when compliance with environmental laws results in maintaining environmental quality; however, despite attempts to quantify the economics associated with improved environmental quality, many businesses perceive the benefit as largely intangible. Furthermore, because organizations such as businesses and municipalities bear most of the direct costs of complying with environmental laws, they often resent the promulgation of stricter and more expensive regulations. Sometimes this resentment is construed as opposition to enhanced environmental quality, although in reality the resentment is directed only toward the costs and other obligations imposed by environmental requirements. For organizations, it is important to recognize a distinction between environmental regulation and environmental quality, as the cost benefits to regulated organizations of total environmental compliance are more easily perceived than the environmental quality benefits. In other words, regulated organizations should go beyond considering the costs of regulatory requirements and consciously account for the benefits of compliance in their business strategies.

The fact that environmental costs are imposed on regulated parties may be largely responsible for the lukewarm or even hostile attitudes that a considerable number of managers show towards environmental laws. However, in many respects, environmental compliance is similar to other business activities that impose costs but do not render benefits capable of simple or rapid measurement. Examples of such activities are preventative maintenance, customer relations, and quality control. Most organizations give these activities considerable attention because they have found that mastering them reduces long-term costs and avoids impediments to operations. Because the benefits are well recognized and considered important, most organizations expend considerable resources on these activities and expect a high level of performance from the employees charged with executing them. However, many organizations do not attribute the same level of importance to environmental compliance as they do to matters such as customer relations because their managers do not fully understand the consequences of noncompliance or the kind of resources they need to ensure compliance. As a result, they miss out on the benefits of total environmental compliance.

Although many managers believe that complying with environmental laws avoids the risk of penalties, the primary benefit for most organizations is retaining control over their operations. The issues associated with environmental noncompliance are complex, and the consequences extend beyond the specter of penalties. For example, courts have broad authority to order compliance with environmental laws and can direct or restrain regulated parties with regard to their principal activities. Similarly, government agencies are aware of this power and often insist on draconian conditions when settling environmental allegations. By ensuring compliance, organizations are able to avoid situations where courts and agencies impose cumbersome solutions that unduly restrict their activities.

Being in compliance with environmental laws has benefits even outside the context of enforcement. Organizations that maintain compliance over the long term need not restrain or misrepresent their activities for fear that their misconduct will be discovered. Such organizations will always be able to lease facilities or obtain government contracts without having to misrepresent their compliance status; likewise, organizations that do not misrepresent their air emissions have no reason to refrain from expanding facilities for fear that prior misrepresentations will be revealed in subsequent applications. Compliance also allows organizations to seek assistance freely from consultants and trade associations without the fear of disclosing violations to outside parties.

The ability to enter into business relationships is also enhanced by ensuring compliance. For instance, compliance facilitates the ability to contract or form partnerships with parties who otherwise would be unwilling to risk liability for another's misconduct. The absence of compliance problems also enhances the ability to interact with other parties in joint ventures, as issues concerning environmental liability are avoided. By addressing environmental matters before opportunities for business relationships occur, organizations can hope to expedite the formation of relationships and make them more productive.

Another benefit of environmental compliance is that it removes a barrier to better employee relations. Generally speaking, employees disapprove of organizations that violate environmental laws, and noncompliance tends to reduce respect for their employers. On the other hand, effective efforts at compliance indicate a sense of institutional responsibility, which is likely to enhance employee loyalty. Organizations that comply with environmental laws also avoid dilemmas created when employees feel pressured to commit violations. In most cases, employees resent having to participate in noncompliant activities, and this is reflected in negative attitudes towards the employer. Furthermore, some employees may feel compelled to inform third parties such as government agencies or environmental groups about environmental misconduct.

1.2 THE RISKS ASSOCIATED WITH NONCOMPLIANCE

The consequences associated with any particular violation will vary depending on the facts of the case and the imagination and zeal of the parties seeking to redress the alleged transgression. Furthermore, the collateral legal liabilities associated with violations are sometimes more complicated than the underlying enforcement issues. It is not unusual for government enforcement actions to be followed by legal actions commenced on behalf of shareholders, environmental groups, employees, unions, and other government agencies. In addition, organizations may find it more difficult to obtain government approvals for activities such as facility expansions and permit renewals. This is not to say that multiple and protracted litigations result every time noncompliance occurs; however, it is important to understand the possible ramifications of noncompliance in order to know what risks and handicaps are incurred by violating environmental laws.

1.2.1 Enforcement Actions

Enforcement actions almost always seek to assess penalties to punish the violator and obtain a court order to force expeditious compliance. In addition, in many cases the parties who file the action seek injunctive relief, which occurs when a court orders the defendant to take action needed to come into compliance. Environmental laws are most often enforced by government agencies, although many laws contain provisions that allow private citizens to litigate enforcement claims when the government declines to do so. Although the citizen suit provisions generally require private parties to seek penalties to be paid to the government, many cases settle on terms by which the alleged violator agrees to provide some private benefit to the plaintiff.

Injunctive relief in the context of enforcement can be very expensive. Courts have the power to order parties to comply with environmental laws and therefore can direct or restrain certain activities. For example, a facility that is violating a permit can be ordered to cease operations or to install additional pollution control equipment. In criminal matters, courts frequently impose conditions of probation such as requiring companies to implement environmental management systems or conduct environmental audits.

Another ramification of enforcement actions is the high cost of defending them. In some cases, the legal fees can exceed the potential penalties. This causes some organizations to settle cases even when they have meritorious defenses. Environmental litigation also requires employees to respond to document requests, attend depositions, and appear at settlement conferences and court hearings. The time and effort required to meet these demands often mean that key employees are unable to devote adequate attention to other responsibilities.

1.2.2 Suspension and Debarment

Organizations that are charged with or convicted of environmental crimes are subject to suspension or debarment from government contracts. Under these programs, the organization will be prohibited from obtaining contracts from the federal government for a period of several years.[1] In many cases, organizations facing suspension or debarment can negotiate agreements to avoid being barred from federal contracting; however, such agreements usually require that the organization maintain a compliance program that emphasizes monitoring and detection for future violations. For example, the agreement may require that managers sign a certification stating that they will not knowingly violate environmental laws and that the company will make compliance audits available to government investigators.

1.2.3 Securities Enforcement

The U.S. Securities and Exchange Commission (SEC) requires publicly traded companies to disclose in periodic reports any judicial or administrative proceedings that relate to the environment and could either have a material adverse effect on their financial condition or involve a claim for penalties that exceeds $100,000. In addition, the costs of complying with environmental laws must be disclosed if they will have a material economic effect on the company.[2] Understating environmental liabilities can give rise to proceedings by the SEC to enforce the disclosure requirements under federal securities law and also provide a basis for private lawsuits by shareholders.

Potential enforcement liabilities may affect how companies publicly characterize their environmental records. The SEC has indicated that companies whose policies or approaches toward environmental compliance are reasonably likely to result in substantial penalties or other significant adverse effects may need to disclose the likelihood and magnitude of the consequences associated with such policies and approaches. In addition, companies have been sued by shareholders for extolling commitments to environmental values that were inconsistent with recent enforcement histories that were only partially disclosed to investors.[3]

1.2.4 Directors and Officers Litigation

Directors and officers have a fiduciary duty to exercise due care in administering the business of a corporation. Although this duty is tempered by a realistic assessment of what directors and officers can reasonably be expected to do in the context of actual business, they can be held liable for failing to respond to obvious problems. Several lawsuits have been filed against directors and officers following media reports of large corporations being subjected to major enforcement actions.[4]

1.2.5 Employee Relations

Environmental noncompliance can profoundly affect the relationship between employers and employees. Most employees care about environmental compliance and resent violations on the part of their employers. Resentment can cause some employees to disclose unlawful conduct to government agencies, environmental groups, or the news media. In addition, several laws protect employees from retaliation for reporting violations to government agencies.[5,6] Not only do these whistleblower statutes expose retaliating employers to liability, they complicate disciplinary issues for any employer with compliance problems, as the employer must be able to prove that any discipline against a whistleblower was made for reasons other than retaliation.

1.2.6 Reputation and Good Will

A company's reputation is a very valuable asset because it facilitates its business activities. Matters involving regulatory agencies tend to proceed more slowly when organizations have established a poor compliance history, as the agency staff tend to be more cautious when evaluating issues. For instance, constructing or expanding manufacturing operations almost always requires that an agency issue some sort of environmental permit, but a bad reputation can bog down the permitting process when agency staff wrestle with how to regulate an entity they distrust. Bad reputations also make it more likely that private citizens will challenge the issuance of new permits.[7]

In addition, a reputation for noncompliance can cost a business some level of sales. Some consumers tend to shy away from purchasing products from companies they disfavor because of environmental issues. Similarly, companies may be reluctant to enter into joint marketing ventures with organizations that are viewed by the public as insensitive to environmental concerns. A poor reputation can also cause commercial purchasers to order from other vendors because of concerns that government restrictions on operations will impair the ability to meet sales orders placed with noncompliant organizations. In fact, there have been instances where salesmen from competing companies have raised such concerns with the customers of companies facing major enforcement actions.

1.2.7 Personal Risks

The consequences of environmental violations also extend to individuals. With regard to legal liability, it has become established government policy when criminal misconduct is at issue to investigate the responsible managers along with the organization;[8] however, individuals can suffer even when they are not subject to legal liability. Environmental matters that result in penalties and other consequences can attract negative attention from senior managers. In addition, the government has successfully pressured many organizations into adopting compliance programs where employee discipline is a prominent feature. Employees in such organizations who are deemed responsible for noncompliance can reasonably expect some sort of adverse action. In many cases, responsibility for or even association with serious, protracted, or repeated noncompliance will damage their career prospects.

1.3 HOW ENVIRONMENTAL LAWS ARE ENFORCED

To understand more fully the ramifications of noncompliance, it is helpful to know how environmental laws are enforced. There are four basic methods by which environmental laws are enforced: (1) administrative actions, (2) civil enforcement by the government, (3) civil enforcement by private parties, and (4) criminal enforcement. Each method presents special issues, but the overall effect is to provide the government and environmental activists an arsenal that allows for practical enforcement of almost any type of violation. For example, the availability of administrative enforcement allows agencies to enforce simple or small matters at less cost than would be required with more formal judicial processes. Civil enforcement allows agencies and activists to avail themselves of judicial procedures that allow for significant discovery of documents and testimony prior to trial. Likewise, criminal enforcement gives the government the ability to impose severe sanctions such as the incarceration of individual wrongdoers.

1.3.1 Administrative Actions

Most federal environmental statutes and many state laws allow government agencies to use administrative actions to enforce environmental laws. Although the respondents in such actions ultimately have the right to some judicial review, most of the proceedings are conducted by a government agency. Agencies may seek the same type of relief as sought in civil actions, but the lower evidentiary and procedural requirements allow administrative actions to proceed faster and at less cost than matters filed in court. In addition, administrative tribunals are often staffed by judges who have special expertise in hearing environmental matters. This expertise helps to facilitate administrative proceedings, as these judges generally do not require as much briefing on the issues as is required in court actions where the judges do not specialize in hearing particular kinds of cases.

Agencies generally use administrative actions to handle routine or straightforward enforcement actions. The U.S. Environmental Protection Agency (EPA) also uses administrative enforcement when it does not want to refer a matter to the U.S. Department of Justice as it generally must do for enforcement matters that will be filed in court. Most violations discovered during the course of government inspections are enforced administratively, as the facts and legal issues are usually straightforward. Some statutes, such as the Toxic Substances Control Act, specify administrative actions as the primary means of enforcement.[9] Other statutes, such as the Clean Air Act and Clean Water Act, give the government discretion in whether to use administrative or judicial processes for enforcement.[10,11]

The primary attractiveness of administrative enforcement to the government is the relative ease with which a matter may be filed and prosecuted. By using administrative enforcement, agencies can avoid extensive discovery efforts and are not as constrained with regard to evidentiary requirements. For example, depositions are rarely taken in administrative proceedings. The downside to administrative enforcement is that the lack of discovery and looser evidentiary standards make it more difficult for parties to prepare cases thoroughly. As a result, administrative hearings frequently incorporate aspects of "trial by ambush" where parties try to gain advantage by presenting surprise evidence.

Administrative decisions may be appealed to courts, but the decisions are usually upheld. Appellate judges will consider only matters covered in the record of the underlying proceeding and generally focus on whether the law was properly interpreted and the proper procedures followed. Parties are almost always barred from rearguing factual issues or raising legal issues that were not presented to the administrative law judge.

1.3.2 Civil Enforcement by the Government

Government agencies tend to favor civil enforcement when cases involve substantial noncompliance, chronic violators, or recalcitrant parties. Because civil actions are filed in courts, the parties must follow strict procedures that govern motions, discovery, trial evidence, and judgments. Although these procedures allow parties to prepare more fully for trial than is possible in administrative actions, they also make the process more expensive, burdensome, and intrusive. Civil cases are commenced by filing a complaint. After the defendant files an answer and the preliminary motions are decided, the parties usually engage in discovery. The right to discover evidence is very broad in civil litigation. For instance, parties may require their opponents to produce documents, answer written questions, and submit witnesses to pretrial questioning. Following discovery and pretrial motions, the court will set a matter for trial if it has not settled. Judgments from trial courts may be appealed, but the review is usually limited to issues of whether the trial court followed the required procedures or misapplied the law. Factual findings are rarely reviewed.

One of the more significant risks faced by noncompliant parties in civil litigation is that discovery will reveal violations that were unknown to the government when the litigation began. Requests for the production of documents can encompass many sources of information, including items such as audit reports, operating logs, and expenditure requests. Such documents are scrutinized for evidence of wrongful acts, damaging inconsistencies, and leads for additional discovery. Depositions may also reveal damaging information and provide leads to other adverse evidence. Even when additional violations are not discovered, the discovery process may reveal information that the government will use to argue for higher penalties. For example, documents generated in the course of requesting an appropriation for pollution control equipment may show that a senior manager delayed the approval of expenditures needed for compliance. Such evidence indicates that the violation was condoned by high-level officials and justifies the award of higher penalties.

1.3.3 Civil Enforcement by Private Parties

Most of the major federal environmental statutes and some state statutes authorize private parties to file lawsuits to address environmental violations. The most common form of private action is the citizen suit in which the plaintiff theoretically steps into the shoes of the government and pursues a court order requiring the alleged violator to pay penalties and comply with the law. Nominally, any penalty so awarded is to be paid into the U.S. Treasury, although in practice most private plaintiffs settle on a basis in which some form of private benefit accrues. In some cases, private citizens may sue violators under bounty hunter and *qui tam* provisions and receive part of the penalty or award assessed by the court.

Congress included the citizen suit provisions in the federal environmental statutes to address concerns that the government does not always diligently enforce environmental laws. These provisions allow private citizens to file enforcement actions when the government has declined to enforce against particular violations. If the government has prosecuted or is diligently prosecuting a civil or criminal action, private citizens are generally precluded from proceeding with citizen suits. Citizen suits are also precluded under some statutes if the alleged violation is already subject to an administrative penalty action. To ensure that the government has the opportunity to commence enforcement, private citizens must notify the government and the alleged violator of the intent to sue prior to filing the suit. Under most circumstances, citizen suits may not commence until a certain period, generally 60 or 90 days, has elapsed after providing notice. If the government commences enforcement within that period, the citizen suit will be barred, although a court may still allow the citizen party to participate in the case in limited ways.

Although nominally the equivalent of government enforcement actions, the practical ramifications of citizen suits often differ.[12] For instance, environmental groups generally do not feel the same constraints with respect to commencing enforcement as some government agencies do. Citizen suits may be motivated by considerations such as publicity, disagreement with regulatory decisions, and the desire to coerce private parties into contributing funds to projects favored by an environmental group. Because the motives of government agencies and citizens can differ, issues such as fairness and reasonableness are less significant in discouraging enforcement actions filed by environmental groups.

Another difference between government enforcement and citizen enforcement is that private parties can generally recover their litigation expenses if they prevail. The ability to recover these costs enables groups to fund litigation that otherwise would not be feasible. In fact, some attorneys have established lucrative practices by specializing in the prosecution of citizen suits. Sometimes, groups are even able to persuade courts to award litigation fees that exceed the actual costs incurred. Because the recovery of litigation expenses is a critical issue to most citizen litigants, citizen suits tend to be directed to industrial and municipal sources even though they may not be the primary sources of pollution in a particular locality.

The False Claims Act is intended to protect the federal government from being defrauded by its contractors but in some cases provides another means of pursuing environmental litigation.[13] Under this statute, government contractors and subcontractors are liable for fraud damages and statutory penalties if they have obtained payments but failed to perform a material requirement of a contract or otherwise avoided an obligation to the government. Because many government contracts require contractors to comply with environmental laws, and sometimes require compliance certifications, environmental noncompliance can give rise to liability under the False Claims Act.

The False Claims Act allows private citizens to sue violators on behalf of the federal government under its *qui tam* provisions. To file such actions, a citizen must serve the government with a copy of the complaint and all material evidence in his or her possession. The government then has 60 days to decide whether to take over the claim. If the government prosecutes the action, the citizen is entitled to receive between 15 and 25% of the treble damages and fines recovered by the government. Should the government decline to prosecute, the citizen may proceed with the claim and will be entitled to recover up to 30% of the judgment and his or her litigation expenses. Recovery is only allowed, however, if the citizen provided information that was not already known to the government or revealed in some public way before the claim was filed.

Some statutes allow private parties to file enforcement actions in which they can recover a percentage of the penalty awarded and their litigation expenses. For example, California's Proposition 65 allows plaintiffs to recover 25% of the penalty awarded by the court.[14] These suits, while generally subject to the requirement of providing prior notice to the government, are easier to commence than *qui tam* actions, because the plaintiff need not possess information unknown to the government or prove that the alleged violator intended to violate the law.

1.3.4 Criminal Enforcement

Criminal enforcement is reserved for egregious violations, although reasonable minds sometimes differ over whether specific kinds of conduct are in fact egregious. The potential for differences of opinion sometimes leads to situations where conduct that is perceived to be innocuous by a regulated party will be viewed as criminal by the government. Nonetheless, the government tends to base decisions on whether to pursue criminal enforcement according to factors such as whether substantial environmental harm occurred and the flagrancy of the violator's conduct;[15] however, there are no bright-line rules for determining when violations will be prosecuted as civil or criminal matters.

Unlike other means of enforcement, the government tends to direct criminal actions at individual employees as well as the entities that employ them. Furthermore, the payment of a penalty by the organization will not generally resolve the issues of individual liability. This makes it difficult and sometimes impossible for organizations to shield their employees from the consequences of criminal conduct. Although some criminal cases have been settled such that the organization agreed to be held liable in exchange for an agreement not to prosecute individual employees, such cases are no longer the norm in environmental matters. This means that in most environmental criminal matters, convicted employees will have to pay fines and often serve time in prison.

Similarly, organizations cannot shield themselves from criminal liability by blaming the conduct on their employees. To convict a corporation, the government need only prove that an employee's criminal act was within his or her scope of employment and committed with the intent to benefit the company. It is no defense that the company policy prohibits such conduct or even that the employee was explicitly instructed not to engage in the unlawful act. If an employee violates an environmental law, the company is usually held fully liable.[16] Moreover, the legal requirements to convict individuals of environmental crimes are frequently less stringent than those for other types of crime. Under a legal principle known as the *responsible corporate officer doctrine*, corporate officials can be convicted of environmental crimes even if the government offers no evidence that they directly participated in or knew of the misconduct. All that the government is required to establish is that a particular corporate official had the authority to prevent or promptly correct the violations and that the corporation allowed the violations to persist.

Although the federal government has rarely prosecuted environmental offenses strictly on the basis of the responsible corporate officer doctrine, it does use theories of liability that are very close to the doctrine.[17] For example, rather than requesting trial courts to instruct juries to convict based merely on a person's position within a company, the U.S. Department of Justice has requested courts to instruct juries that they may consider whether a defendant's knowledge of unlawful activity can be inferred from the position in the company. This instruction authorizes juries to convict if they believe that defendants would have known about violations because of their status within the company.

Criminal matters proceed somewhat differently than civil and administrative matters. Most criminal matters are initially investigated informally by government agents. During the course of such investigations, agents may interview employees during their off-work hours or conduct surveillance of a facility from offsite. In many instances, the target of the investigation will not know that it is being investigated. If the government believes that sufficient evidence of criminal activity exists, the prosecuting attorneys will convene a grand jury to further investigate the matter. The grand jury process is a very powerful investigative tool, as the government can subpoena documents and witnesses for examination behind closed doors. Because grand jury proceedings are secret, the targets of the investigation cannot rebut the evidence presented against them and are not entitled to see it unless indicted. Furthermore, the government need not satisfy any particular level of proof to obtain an indictment. It is thus not surprising that grand juries almost always follow the prosecutors' recommendation to issue indictments.

Another investigative tool used by investigators and prosecutors is the search warrant. Once the government establishes that probable cause exists to determine that a crime has occurred, it can obtain a search warrant from a judicial officer allowing them to enter onto private property to seize evidence. Typically, while on the premises, government investigators will also attempt to interview employees. These interviews can result in the discovery of damaging evidence, because in many instances the employees will be unfamiliar with their legal rights and may feel strong psychological pressure to speak to the investigators to attempt to clear themselves from criminal liability. It is not unusual for misrepresentations made during such interviews to constitute the basis for the separate and often more serious federal offense of making false statements to a government agency.

After the investigation is completed and the grand jury has issued the indictment, the court will set the matter for trial. In most cases, the government and the accused parties enter into plea agreements and avoid trying the matter. Such agreements provide the parties with some degree of control over the outcome but still result in convictions. If a plea agreement is not reached, the government must prove guilt beyond a reasonable doubt. Although reasonable doubt is a high standard of proof, the actual burden on the government for many environmental crimes is not that great. Most of the criminal provisions in environmental statutes require only that the government show that the defendant knew that he or she was committing the act that violated the law. The government rarely needs to prove in an environmental crimes case that the defendant knew that the act was illegal. In other words, for most environmental crimes ignorance of the law is no excuse.

Case Study: *United States v. Weitzenhoff*, 35 F.3d 1275 (9th Cir. 1993)

Facts

This criminal case arose in Hawaii and involved two managers accused of ordering treatment plant workers to discharge sludge directly into an ocean outfall in order to avoid hauling it to a separate treatment works. Evidence put on by the government indicated that sludge had been discharged from storage tanks directly into the outfall on about 40 occasions in a manner that bypassed the effluent sampler. The government further asserted that most of the discharges occurred at night, the defendants had repeatedly lied to state inspectors about the discharges, and most of the discharges exceeded the effluent limits in the permit.

In their defense, the defendants presented evidence that they believed the discharges were needed to maintain the biological balance of the treatment system and were thus authorized by a provision in the permit that allowed the plant to bypass the treatment systems when necessary for essential maintenance to ensure efficient operation. Although the defendants' efforts to conceal the discharges of sludge tended to show they knew they were violating the law, the government persuaded the court that it needed to prove only that the defendants knew they were ordering the discharge of sludge and that proof of unlawful intent was unnecessary. The judge agreed, and the case was submitted to the jury with the instruction that "the government is not required to prove that the defendant knew that his act or omissions were unlawful."

Following their conviction, the trial court sentenced the two managers to 21 months and 33 months of imprisonment. They appealed the case to the U.S. Court of Appeals for the Ninth Circuit.

Discussion

The Ninth Circuit ruled that even had the defendants misunderstood or misinterpreted the provisions of a wastewater discharge permit, environmental laws constitute public welfare offenses for which Congress intended to place the burden on the defendant to ascertain whether his or her conduct violates the law. Under this reasoning, it is irrelevant whether an individual who engages in conduct that violates a permit requirement knows that the conduct is subject to a particular legal requirement.

Following the decision in *Weitzenhoff*, the regulated community expressed grave concern that the Ninth Circuit had established a framework by which individuals could be convicted of felonies for unknowing violations of permits and similar legal requirements. By removing the requirement that the government must prove that defendants knew they were violating the law, even persons acting in good faith and motivated by the best of intentions could be convicted once the government showed that their acts somehow violated an environmental law. Considering the fact that courts generally defer to the government's interpretation of environmental laws, and the myriad of laws in existence, it is not farfetched to speculate that individuals may be found criminally liable for conduct they believed was lawful but which varied from the interpretation of a government official. Nonetheless, the U.S. Supreme Court declined to consider the *Weitzenhoff* case further, and other federal circuits have since adopted the Ninth Circuit's standard of culpability in environmental criminal cases.

1.4 VIOLATIONS AND ENFORCEMENT DECISIONS

No enforcement action can commence unless the government or other potential litigant has reason to believe that an environmental violation has occurred; however, the pervasiveness, complexity, and documentation requirements associated with environmental laws make it likely that violations will eventually make themselves known. There are many ways in which parties with an interest in commencing enforcement actions come to learn about violations. Government inspections are an obvious means, as is review of the reports and correspondence that regulated parties are required to routinely submit to regulatory agencies. Some statutes and permits require regulated parties to report certain kinds of spills and violations to government agencies. There have been many instances where disgruntled employees have reported misconduct, and unions have also been known to use environmental laws to gain leverage against employers they disfavor.[18] In addition, private citizens have been known to report environmental problems that ultimately led to the discovery of noncompliance.[19]

The discovery of a violation does not necessarily mean that enforcement will result. In some cases, government agencies may tell the violator to correct the problem or even issue an administrative order setting forth a schedule for corrective action. Even when enforcement is seriously considered, agencies generally follow formal processes before commencing the proceeding. However, alleged violators have little control over how enforcement will proceed, and agencies have enormous discretion over whether to file administrative, civil, or criminal actions to enforce environmental laws. Although no firm rules exist with respect to how the government decides to enforce against particular violations, some general principles apply.

How a matter comes to be discovered can make a difference in the way the agency handles it. Matters that are voluntarily disclosed tend to be treated less harshly than those that involve affirmative concealment. Also, violations reported by whistleblower employees or private citizens are sometimes enforced more severely if they present facts that indicate the government would not have discovered them otherwise.

The nature of the violation can affect the likelihood of enforcement. Conduct that affects the integrity of a regulatory program is frequently enforced. For example, failure to keep the proper records or to submit reports in a timely manner often results in enforcement actions even though the violations did not result in environmental harm. Serious misconduct, such as tampering with monitoring equipment or falsifying laboratory data, are usually handled as criminal matters. Environmental harm in itself is rarely a significant factor in decisions to pursue enforcement absent massive environmental damage such as occurs from large oil spills; however, agencies will sometimes target operations in degraded areas or those associated with large releases of pollutants as part of enforcement initiatives.

Another factor affecting enforcement decisions is the attitude of an alleged violator. Government agencies often act on the discovery that a violator has disregarded known problems by resorting to stronger enforcement measures.[15] For example, a company that learns of a violation through an environmental audit and fails to promptly correct the problem should not expect lenient treatment. Violators with good compliance histories will likely face less severe enforcement than those with poor records. Pervasive noncompliance throughout a company or repeated violations of a particular regulatory program may favor severe enforcement such as criminal proceedings, particularly when the government believes that civil enforcement has been ineffective in making the violator understand that compliance is mandatory. A poor history of compliance also creates the appearance that noncompliance is condoned by management and has become part of the corporate culture. In such cases, the government may devote significant resources to investigating whether senior managers should be held personally liable.

Private entities such as activist groups tend to show less restraint than government agencies when it comes to enforcement decisions. A primary concern in private enforcement actions is whether litigation expenses can be recovered, as most groups do not have the resources to fund enforcement activities absent such recovery. Courts normally award these expenses only to groups that prevail in the litigation, thus creating an incentive to pursue matters that they are unlikely to lose. Unfortunately for many regulated parties, it is not particularly difficult for environmental groups to find matters in which liability is easy to prove. Many regulated parties are required to report violations by such means as discharge monitoring reports and compliance certifications. Many environmental groups review agency files pursuant to public-access-to-document laws and find matters in which some degree of liability has already been admitted by potential defendants.

1.5 HOW ENVIRONMENTAL LAWSUITS ARE RESOLVED

Only a small portion of enforcement actions are resolved by decisions made by judges and hearing officers. Most settle by mutual agreement between the parties for reasons such as control over the outcome and the desire to minimize litigation expenses. Nonetheless, settlements are strongly influenced by the sanctions that could be imposed by administrative and judicial tribunals if the matters were actually tried. Although courts and administrative tribunals have flexibility in crafting sanctions, their authority is constrained by civil penalty factors and criminal sentencing guidelines. These guidelines are important not only for the role they play in settling environmental litigation but also because they can affect how organizations respond to the discovery of noncompliance.

Most agencies have prepared penalty policies which establish specific frameworks by which penalties are determined. For instance, EPA penalty policies encompass the components of economic benefit and gravity.[20] The economic benefit component is intended to recover those costs saved by virtue of a party delaying compliance with an environmental law. The gravity component imposes the punitive sanction intended to deter future violations. The effect of these penalty policies is to diminish the discretion exercised by administrative tribunals when assessing penalties. Courts are generally required to follow statutory penalty guidelines and are not strictly bound to penalty policies drafted by agencies; however, most courts rely heavily on these policies when awarding penalties. Because the policies are drafted from the perspective of agencies, many regulated entities view them as harsh and inequitable. Nonetheless, the policies diminish the discretion of judges and make it difficult for violators to argue against the assessment of penalties that appear unreasonably high for matters involving unintentional or technical violations.

The federal government and many states have promulgated sentencing guidelines that severely limit the discretion of judges when imposing sanctions on parties convicted of crimes.[21] The result is that persons convicted of felonies can expect to be incarcerated irrespective of their prior record. The sentencing guidelines also apply to organizations but consider additional factors such as internal efforts to detect or prevent criminal conduct. When sentencing organizations, judges must enhance the fines if aggravating factors were associated with the offense. Among the aggravating factors considered by judges are (1) involvement by management in the offense, (2) prior enforcement history, (3) violation of an order, and (4) concealment of the crime. Judges are also allowed to mitigate part of the fine if the organization had implemented an effective compliance program at the time of the offense, disclosed the misconduct to the government, and cooperated with the government investigation. The organizational guidelines thus give organizations an incentive to disclose criminal misconduct in ways that increase the exposure of their employees to criminal liability. As a result, employees who intended their misguided activities to benefit their employers will find themselves under attack by both the employer and the government.

The government has three processes by which organizations and individuals associated with environmental crimes may be precluded from federal contracting.[22] Under the Clean Air Act and Clean Water Act, convicted entities are automatically placed on the EPA's List of Violating Facilities. Once listed, the violating facility is barred from future contracts or subcontracts with the federal government until the EPA removes it from the list. In addition, upon a determination that the facility is engaged in continued or recurring noncompliance, the EPA has the discretion to list facilities for civil and administrative violations. Suspension is a process that renders parties temporarily ineligible for new contracts or subcontracts pending a final determination of whether a person or organization is sufficiently responsible to do business with the federal government. The EPA often suspends parties shortly after an indictment for environmental crimes. Debarment proceedings are commenced for the same reasons as suspensions but allow the subjected party prior notice and an opportunity to oppose the debarment. Debarments remain in effect longer than suspensions, typically for up to 5 years. The effect of debarment is the same as for listings and suspensions; the debarred party is ineligible to receive federal contracts or subcontracts.

Although it is possible to avoid listings, suspensions, and debarments by negotiating agreements with government agencies, organizations typically have little negotiating leverage and will be forced to accept harsh conditions. At a minimum, organizations will be required to implement a compliance program to ensure a high level of business integrity. Such programs generally incorporate elements such as corporate policies, auditing, and training requirements. The government may also impose onerous terms that effectively allow agencies to manage important aspects of the organization's activities. For example, an agreement may require the organization to dismiss or reassign certain employees, report future violations, and allow the government access to sensitive and confidential information. Agreements may also require employees to sign affidavits attesting that they will not participate in violations of environmental laws. Organizations that enter into these agreements find they can adversely affect employee relationships and greatly reduce their ability to protect themselves from future or collateral legal liability.

1.6 HOW SETTLEMENT CONSIDERATIONS AFFECT OUTCOMES

Environmental enforcement can present organizations with complex issues with ramifications that go far beyond assessing the legal merits of the claim. These issues include the cost of defense, the risk of an onerous court decision, control over publicity, potential discovery of other misconduct, and collateral litigation by shareholders or employees. Many decisions to settle enforcement actions are thus dominated by factors other than the merits of the case.

The primary incentive to settle enforcement litigation for most parties is to have some level of control over the result. Although they may not necessarily be satisfied with the terms of settlement, in many instances the parties prefer negotiating a result they know they can live with to risking that a court will render a devastating award. Control over the outcome can be very important in environmental matters because of the uncertainty associated with what a judge may decide. The penalties awarded for environmental violations can be enormous, and courts have vast discretion in determining how to order facilities to come into compliance. For example, a party may prefer to settle on the basis of installing additional pollution control equipment rather than risk that the judge will order it to shut down a production facility. In addition, parties can use settlements to resolve matters in ways that fall outside the scope of what a court could order. For instance, agencies are often willing to negotiate lower penalties in exchange for an agreement by a violator to perform an environmentally beneficial project that is not required by law.[23]

Another important reason why parties settle is to avoid the high costs of litigating enforcement actions. Enforcement litigation is expensive and usually requires employees to spend copious amounts of time responding to litigation matters which precludes them from working on more productive tasks. As a result, many organizations decide that it makes sense from a business perspective to minimize their losses by settling the matter.

Sometimes organizations will settle matters to prevent adverse parties from discovering other violations. The discovery processes in civil and criminal litigation can be very intrusive and often extend beyond the matters covered by the claims. Parties in environmental litigation may expect that their documents, employees, and consultants will be thoroughly examined by the opposing parties. Should these examinations reveal that other serious violations have occurred, it is likely that the litigation will escalate. Faced with the possibility that other misconduct will be revealed during discovery, it is understandable that some organizations will want to settle to avoid uncomfortable disclosures.

Issues that are collateral to the litigation can encourage parties to settle. For example, by settling on terms that avoid admitting liability, organizations can make themselves less vulnerable to other lawsuits such as securities litigation and shareholder claims. Similarly, organizations that need to obtain permits may want to settle enforcement matters to reduce the level of conflict with a regulatory agency.

The fact that the merits of a case do not always dominate the considerations involved in settlement means that favorable outcomes can be difficult to achieve. As discussed previously, enforcement litigation places a much larger burden on organizations than the risk of penalties. Although organizations can be forced by circumstances outside their control to litigate these matters, their best opportunity to maintain control over their operations is to adopt sound environmental compliance practices to reduce the prospect of enforcement. It is conceded that compliance cannot prevent parties from filing frivolous lawsuits; however, it is by far the most effective means of protecting organizations from the consequences of noncompliance.

2 Legal Liability and Attitudes Toward Compliance

Most environmental professionals understand that organizations and individuals can be held liable for environmental noncompliance, but many do not fully appreciate the legal bases for imposing liability. In general, the imposition of liability in environmental matters depends on factors such as culpability and the protection of public interests. However, legislative concerns about protecting human health and the environment have resulted in a general lowering of the culpability needed to impose liability for environmental noncompliance relative to other forms of liability. By understanding the basic principles by which liability is imposed, environmental professionals can increase their sensitivity to compliance issues and become more proficient at avoiding noncompliance.

Another factor affecting compliance is the attitude that organizations and individuals exhibit with regard to environmental laws. Although there is a general awareness of that poor attitudes increase the likelihood of noncompliance, relatively few environmental professionals have considered the obstacles that specific attitudes pose to environmental compliance and how to overcome them. As is the case with understanding legal liability, becoming familiar with how attitudes affect compliance should make any environmental professional more proficient at avoiding violations.

2.1 LEGAL LIABILITY

Legal liability depends on two elements: (1) the occurrence of an act prohibited by law, and (2) the intent of the party who committed the act. Absent the prohibited act, there can be no legal liability; likewise, a person who does not show the requisite degree of intent cannot be held liable for an offense that requires a particular mental state. An aspect of legal liability for noncompliance that every environmental professional should understand is that most environmental laws do not require that the conduct affect the environment. This means that parties can be held liable for acts that violate environmental laws regardless of whether they harmed or were intended to harm the environment.[24]

The lowest standard of legal liability is known as *strict liability*. To be found liable under this standard, it is necessary only to have committed the prohibited act, and no affirmative mental state must be proven. This means good faith and ignorance of legal requirements do not preclude liability for strict liability offenses. In most cases, absent extreme coercion, the occurrence of the act will be sufficient to sustain a finding of liability. What is critical for the environmental professional to understand is that most environmental laws impose strict liability on those who violate them. For example, a large-quantity generator who accumulates hazardous waste without a permit for more than 90 days violates the Resource Conservation and Recovery Act (RCRA), irrespective of whether the generator believed in good faith that the waste was stored for less than 90 days and whether any harm occurred. The only relevant issue to finding liability is whether or not the waste was stored past the allowable time.

Negligence requires a finding that the person or organization did not use the degree of care expected from a reasonably prudent person under the circumstances. Because negligence requires an evaluation of whether the acts were imprudent, it is a higher standard than strict liability. The application of the negligence standard to environmental laws is usually limited to misdemeanor criminal offenses. To illustrate the difference between negligence and strict liability, one could look at an example of a laboratory manager using outdated reagents to analyze wastewater samples which results in the manager's employer misreporting the levels of pollutants to the agency that administers the permit. The employer would be held strictly liable for the inaccurate reporting regardless of the manager's reasons for using outdated reagents, as the standard for civil liability under the Clean Water Act does not require the government to show the intent underlying the violation.[11] However, the manager could only be convicted of a criminal offense that requires negligence if the government proves that the manager had reason to expect that the reagents were outdated. For instance, if the manager were new and had been told by his predecessor that the reagents were fresh, it would be unlikely that a jury would find that the manager acted imprudently and any claim of negligence would fail.

The highest level of intent is required for offenses that are characterized in statutes as *willful or knowing violations*. Willful or knowing acts are those committed either intentionally or with conscious disregard of the consequences. Some statutes require the government to prove that the violator not only knew that he or she was committing the prohibited act but also knew that the act was unlawful. These kinds of violations are characterized as *specific intent offenses*. Other statutes require the government to prove only that the violator intentionally committed the wrongful acts and do not require proof that the violator knew they were unlawful. These kinds of violations are considered *general intent offenses*. Many environmental statutes that prohibit knowing violations do not clearly state whether they are specific or general intent offenses; however, courts have construed most of these statutes to be general intent offenses.[25] Thus, for most environmental crimes, the government need not prove that the defendants intended to violate the law but only that they knew they were committing the relevant act.

2.1.1 Individual Liability

The most obvious way an individual can face liability for a criminal offense is to directly participate in or authorize its commission. If an employee falsifies monitoring data, lies to a government investigator, or orders hazardous wastes to be illicitly disposed, he or she can be held legally accountable regardless of whether the intended beneficiary of the crime was the employer. Furthermore, employees can and have been held liable for committing crimes that were ordered by superiors.[25] It is no defense that a participant would have been fired or disciplined had they refused to follow such orders. In fact, the government often pursues lower level employees as targets of criminal investigations to get them to testify against their superiors in exchange for lesser sentences.

Individuals can also be liable for acts in which they did not directly participate if they assist others in committing a crime or otherwise further its commission. For example, a person who collects and dilutes a wastewater sample knowing that others will report the erroneous result to the government can be criminally liable for making a false statement even though he or she did not actually prepare the monitoring reports that were submitted to the government. By committing acts that cause a crime to be committed, a person can be held liable for aiding and abetting that crime.

The *responsible corporate officer doctrine* is another means for imposing liability on individuals who do not directly participate in criminal acts that violate public welfare laws such as environmental statutes. This form of liability is based on the principle that it is proper to hold people criminally liable for a public welfare offense when they have a responsible share in the furtherance of the transaction which the statute prohibits. Under this doctrine, senior officials may be held personally liable if they had the responsibility and authority to have either prevented or promptly corrected a violation but failed to do so.[26] In other words, it is possible for managers to be held criminally liable by virtue of their position within an organization.

The federal government rarely prosecutes environmental offenses strictly on the basis of the responsible corporate officer doctrine but does use similar theories of liability. For example, the U.S. Department of Justice has argued in court that juries may find a defendant guilty if they believe that he or she would have known about the unlawful activity because of their position and responsibilities in the company.[17] By using this tactic, the prosecution can avoid having to produce direct evidence that a senior manager knew of misconduct.

Another means by which managers can be held individually liable when they claim they were unaware that unlawful activity was occurring is the doctrine of *conscious avoidance*. Under this doctrine, willfully avoiding knowledge about wrongful conduct can serve as the basis for imposing legal liability. Conscious avoidance is typically evidenced by a manager instructing subordinates that he or she does not want to know about wrongful conduct or compliance problems.[27] The motive behind conscious avoidance is generally to have subordinates commit violations while attempting to maintain plausible deniability in the event the misconduct is discovered by government regulators. However, under the doctrine of conscious avoidance, defendants can be held liable for being aware that misconduct was probably occurring and failing to investigate.

One of the significant ramifications associated with the legal doctrines that allow individuals to be convicted based on indirect participation is that they reduce the evidentiary burden placed on the government. Instead of having to prove that a manager actually knew or ordered misconduct, the government can rely on lesser evidence to sustain its burden of proof. These evidentiary requirements are also much easier to present. For instance, a particular manager's position and responsibilities within an organization will generally not be contested in a trial. By showing that the manager could be expected to have known of the misconduct because of his role in the organization, the government can ask the jury to draw the inference that the manager either knew of the misconduct or willfully avoided acquiring such knowledge.

2.1.2 Corporate Liability

Organizations such as corporations and municipalities can be held both civilly and criminally liable for environmental violations. Under the doctrine of *respondeat superior*, an employer is liable for the acts of employees performed within the scope of their employment.[16] Management involvement in the wrongdoing is not required for organizations to be held liable. Under this principle, organizations can be held liable even for the acts of low-level employees irrespective of whether these employees acted contrary to company policy or even against directives from superiors. Similarly, the imposition of liability does not depend on whether the organization actually benefited from the misconduct so long as it was intended to confer at least some incidental benefit.

In addition to being held liable for the misconduct of an individual employee, organizations can be held liable on the basis of information known among several employees although it was not known by any single individual who comprehended its significance. The doctrine of *collective knowledge* is important in the environmental context because duties affecting the environment tend to be compartmentalized, and the associated legal requirements may be obscure or complex. Thus, organizations can be held liable for false reporting in cases such as when an employee falsifies data that he or she does not know will be reported to the government and another employee reports the data to the government not knowing that they are false.

2.2 ATTITUDES AND THE ROAD TO NONCOMPLIANCE

The road to noncompliance, if not paved with the attitudes of those who violate laws, is at least constructed on top of them. Attitudes that can lead to noncompliance reflect a broad range of motives and beliefs. Many people have positive feelings toward citizenship and environmental quality which are reflected in their attitudes toward environmental compliance. Some people have attitudes that are less likely to result in consistent compliance, and sometimes these attitudes are rooted in a misunderstanding of how environmental laws are applied to regulated entities. There are also cases when a negative attitude toward environmental compliance reflects a conscious disregard of societal obligations. In any case, the ability to recognize and understand attitudes toward environmental compliance is important if widespread noncompliance is to be prevented. For example, resolving problems caused by ignorance will require approaches that differ from those used to resolve problems caused by greed.

A major cause of poor attitudes towards compliance is uncertainty. A common source of uncertainty is the applicability of legal requirements to regulated activities. Employees often misunderstand their obligations to comply with environmental laws because the regulations are difficult to comprehend or because they misread the potential enforcement consequences. For instance, people who have not experienced significant enforcement activity tend to become more complacent about violations than those parties who have had to contend with the consequences associated with violations. Similarly, many people misunderstand the stringency of environmental requirements and perceive wrongly that substantial compliance is sufficient to avoid enforcement and other liabilities. In addition, changing attitudes of government agencies about the use of enforcement mechanisms have caused many regulatory parties to form misconceptions about the likelihood of whether particular violations will or will not be enforced.

Of equal importance to regulatory uncertainty is the lack of ethical certainty that many organizations and employees feel when it comes to complying with environmental requirements. Unlike regulatory uncertainty, which is influenced mostly by external factors, organizations have a great deal of control over how they fashion their internal ethic. Unfortunately, many organizations do not avail themselves of this control and fail to develop effective and meaningful guidelines regarding their manner of operation. Instead, the collective ethic evolves in response to individual events and the reactions thereto and tends to reflect the diverse attitudes and interests of individual managers rather than an organizational directive. For some organizations, this results in a positive ethic that values honesty and responsibility, while in others the result is a negative ethic fostered by a history of shortsighted but pragmatic solutions to problems for which the informal ethical principles failed to provide employees with clear or sufficient guidance on how to act.

Even though the roots of some attitudes can be traced to uncertainties over how entities are governed by themselves and others, it would be foolish to ignore the fact that some people have attitudes that reflect personal shortcomings. Although organizations desire employees who are diligent, careful, and responsible, the reality is that most organizations have to deal at least occasionally with employees who are lazy, sloppy, or lack commitment. These shortcomings not only express themselves in the actions of individual employees but may also show up institutionally, such as when managers defer responsibility for misconduct to subordinates and vice versa.

To successfully overcome the problems attributable to attitudes, environmental professionals need to be able to deal with the individual and organizational issues presented by the effect that attitudes have on achieving environmental compliance.

2.2.1 Attitudes Based on Ethical Shortcomings

Although attitudes based on ethical shortcomings are responsible for a small portion of compliance problems, they frequently are at the root of the most egregious violations. Furthermore, the government tends to enforce violations that reflect unethical attitudes more severely than other kinds. For example, the intentional bypassing of treatment equipment to save costs is more likely to be prosecuted criminally than an accidental spill caused by equipment failure. Compliance problems caused by senior managers with ethical shortcomings can present a serious problem because they tend to become widespread and involve misconduct which may entail severe penalties; however, the ethical shortcomings of low-level managers and even hourly employees can also expose organizations to significant consequences, especially when the degree of management oversight indicates complacency or tolerance on the part of supervisors.

2.2.1.1 Greed

Greed in the organizational context may be defined as an excessive desire for financial performance. As such, greed should not be confused with having a profit motive or maximizing the use of financial resources, as a healthy appreciation of finances usually encourages a proper level of attention to compliance matters. Nonetheless, the temptation to increase financial performance by avoiding necessary environmental expenditures is something that organizations need to understand and counteract if they are to avoid costly penalties. It is also important to understand the mechanisms by which greed leads to noncompliance. Flagrant measures to cut costs, such as illicit dumping of hazardous wastes and operating processes without the required pollution control equipment, are obvious examples, but there are more subtle manifestations of greed that can nonetheless lead to serious noncompliance. One of these is providing a level of environmental control that is sufficient most of the time but woefully inadequate at critical moments. For example, a company may decide that the costs associated with sizing a treatment unit to handle loadings from infrequent events are not economically justified, thus they incur infrequent but serious violations. Another common manifestation of greed is the decision to delay environmental expenditures. For instance, by delaying the installation of equipment past a regulatory deadline, an organization may hope to meet some other financial objective at the expense of temporary but deliberate noncompliance.

To curb greed as a root cause of noncompliance, organizations should address the motivations behind forgoing compliance to incur a financial benefit. At the organizational level, the issue of financial performance should be presented in terms of how the organization will benefit economically as a result of particular conduct or decisions. Because decisions in this context usually do not pose direct financial benefits to employees, their conduct is amenable to being shaped by implicit or explicit corporate policy; therefore, senior management needs to communicate to subordinate managers that the organization expects that all activities either should be in environmental compliance or should not be operating.

Management also needs to address the motivations that individual employees may have to violate environmental laws in order to incur a personal benefit. Some organizations unintentionally encourage individual managers and employees to evade compliance requirements by making compensation dependent on factors that sometimes conflict with environmental compliance. For example, the organization may consider quarterly performance criteria as a factor in determining the compensation of managers. This practice can encourage managers to act irresponsibly because it tends to focus on short-term gains at the possible expense of higher costs later. Furthermore, it can cause noncompliance to escalate over time as managers take bolder measures to defer environmental expenditures to meet performance criteria. Some forms of employee compensation also encourage noncompliance. For instance, some facilities have set up compensation programs where hourly employees are paid bonuses for meeting production targets. Not only do such programs tempt hourly employees to disregard environmental obligations, they also can place undue pressure on individual employees whose actions can affect whether an entire facility receives a bonus.

Organizations need not forgo compensation based on individual or collective performance but should be careful to ensure that managers and employees understand that short-term financial performance is not the dominant criterion by which their performance is evaluated. For example, organizations should consider a manager's use of judgment and whether his or her actions have exposed the organization to legal risk. Likewise, managers should be evaluated on their ability to forecast costs and to budget their operations in a way that ensures compliance. Similarly, compensation systems for hourly employees should be structured in ways that do not provide unintended incentives to forgo compliance.

2.2.1.2 Contempt for the Law

Some people cannot tolerate being told to do something and sometimes express themselves by showing contempt for the law. This contemptuous attitude is particularly evident when people make known their awareness of a legal requirement and heatedly assert they are somehow above having to comply. In the worst cases, contempt is the only motive for violating laws. In other cases, the attitude will express itself by maintaining that other activities are more important than complying with an environmental requirement.

Contempt for environmental requirements, when it manifests itself, is one of the most dangerous from the enforcement perspective, as violations that result from contempt of the law tend to be blatant and frequently are severe. Violations that stem from arrogance also tend to be addressed with criminal sanctions. In part, the heavy-handed response to arrogant violators may be rooted in the government's belief that its view of the world is the correct one and the high need of some government officials for others to recognize their importance.

Arrogance is a particularly serious problem when manifested in senior management, as it tends to lead to a corporate culture that encourages flagrant violations. One reason is that arrogance on the part of senior managers has a tendency to cause managers at lower levels to adopt the values implicitly communicated to them. As a result, decisions that result in noncompliance come to be made at all levels of the organization and will frustrate systematic efforts to sustain or improve compliance. Another reason is that the behavior of arrogant managers tends to cause subordinates to agree to unlawful conduct to avoid being demeaned.

People who have contempt for the law tend to believe strongly that their view of the world is correct and have a high need for others to agree with their point of view. This can make dealing with such people difficult and frustrating. Sometimes it is possible to address contempt for the law by the way issues involving compliance are structured. If the person resents regulation but is otherwise analytical and intelligent, framing issues in terms of alternative choices can sometimes facilitate good decisions. For example, instead of stating that a proposed facility will require a wastewater treatment unit, the issue can be presented in terms of whether the project should omit the treatment unit, in which case the organization will have to deal with regulators and attorneys, or whether unprofitable regulatory headaches can be avoided by installing the treatment unit. By allowing persons who resent environmental laws to participate in the decision process and exercise judgment with respect to how resources are allocated, they may be more inclined to focus on the needs of the organization instead of their need to combat authority.

Unfortunately, not everyone who has contempt for environmental laws is analytical or even intelligent. In such cases, interpersonal skills will probably determine the time and degree of success associated with decisions involving compliance. An approach that is often productive is to provide information in small doses over time. For example, information that a permit will be required can be provided to introduce the idea that a project will involve environmental obligations. This introduction to the issue can be followed at a later time by information that a treatment system will be required. By providing information in incremental units, one can provide the person some time in which to reflect on the environmental requirements and to accept them psychologically.

Although it is possible to deal with contemptuous attitudes, organizations should nonetheless avoid placing people who do not respect the law in positions where their decisions can affect environmental compliance. Even when their contemptuousness does not result in violations, the effort required of subordinates and peers to deal with compliance issues can create undue strain within the organization. Furthermore, the government takes a dim view of organizations that assign people who are predisposed to violating the law to sensitive positions. For instance, the criteria for effective corporate compliance programs set forth in the *United States Sentencing Guidelines* requires companies to take measures to ensure that employees who are inclined to break the law are not given inappropriate responsibilities.[28]

2.2.2 Attitudes Based on Erroneous Perceptions

Erroneous perceptions about environmental requirements are probably the most significant factor leading to noncompliance with environmental laws. Although most people understand that activities such as illicit dumping and tampering with monitoring devices can result in severe penalties, many do not appreciate how other kinds of violations expose organizations and their employees to liability. One reason for widespread misconceptions about environmental noncompliance is that the imposition of environmental liability falls outside the common experience of most people. Because they are not familiar with the framework in which environmental laws are administered and enforced, they assume that the framework is analogous to other areas in which activities are regulated.

The problem with attitudes that reflect an inaccurate perspective regarding compliance is that they can lead to long-standing and pervasive noncompliance that can result in substantial liability when addressed by the government. Although violations that result from a loss of perspective tend to be less egregious and involve lower levels of culpability than those motivated by greed, their collective effect can hamper operations when deficiencies in regulatory documents, such as permits, become apparent. Furthermore, even minor violations can become grounds for enhanced enforcement efforts when the government views them as being pervasive and historically condoned by management.

Overcoming attitudes based on erroneous perceptions is usually easier than is the case for attitudes based on ethical shortcomings. In many instances, these attitudes are not motivated by the desire to avoid costs or the satisfaction of flouting authority but result from simple misunderstandings. Organizations can go a long way to dispel such attitudes by making compliance an explicit objective as opposed to relying on a tacit understanding that the organization wants to avoid enforcement liability. It is sometimes possible to correct erroneous perceptions, but it is not unusual for people with ingrained attitudes to refuse to accept such instruction. Absent some ethical deficiency, though, most employees will not resist environmental compliance when properly directed by management.

2.2.2.1 Inaccurate Perceptions of Agency Priorities

In general, people have a tendency to attribute importance to matters that bear a tangible connection to something they value. For example, polychlorinated biphenyls (PCBs) are toxic substances, and most people understand that disposing of these substances in unlined pits would likely result in enforcement because environmental quality is something that is valued. One the other hand, not everyone considers information such as the annual document logs (which facilities with equipment containing polychlorinated biphenyls are required to keep) to be particularly important, as the documents do not directly protect environmental quality. As a result, they are often surprised when the U.S. Environmental Protection Agency (EPA) assesses large penalties for missing annual document logs, primarily because they do not understand that the agency considers compliance with its regulatory mechanisms to be a matter of great importance.[29]

Not understanding the basis by which agencies implement environmental laws can result in the "no harm, no foul" attitude in which managers and other employees assume that agencies will require compliance only if an activity has some perceivable effect on environmental quality. Underlying this attitude is the belief that enforcement is directed only at curtailing activities that have a substantial effect on the environment and that government agencies will not be concerned otherwise. Instead of attempting to achieve total environmental compliance, employees with this attitude often favor a practical approach when addressing the environmental aspects of the activities for which they are responsible. For instance, this attitude may manifest itself in managers who consider regulations and permits as general guidelines and not as strict requirements. Unfortunately, the propensity for environmental harm is neither a legal or practical prerequisite for enforcement.[30]

2.2.2.2 Misunderstanding Enforcement Practices

Many people believe that agencies rarely enforce environmental laws or they always give violators a chance to correct noncompliance before commencing enforcement. Although agencies sometimes do decline to enforce against some violations and sometimes do give parties an opportunity to correct violations before assessing penalties, they are not required to do so.[31] Furthermore, although agencies differ in their proclivity towards enforcement, many environmental violations are capable of being enforced by other agencies or private citizens. State agencies are sometimes accused of failing to diligently enforce major federal environmental laws, which has resulted in the federal government or private activist groups stepping in with zealously litigated enforcement actions.[32] Because statutes of limitations associated with environmental laws typically run for 5 years or more, violators that have acted on the perception that the laws would not be enforced can suddenly find themselves subject to hefty penalties.

An important factor at some facilities that contributes to employee misunderstanding of enforcement practices is the presence of activities that are patently out of compliance. After such practices have continued for a period of time, the lack of consequences causes the employees to feel more comfortable about them and to assume that enforcement is unlikely. Sometimes inaction or perceived inaction by regulatory agencies will foster misunderstandings, such as when agency inspectors fail to comment on noncompliant practices or declare a lack of interest in requiring compliance. Other common causes of misunderstanding enforcement possibilities are the perceptions that everyone else in the industry is similarly out of compliance and that agencies are unlikely to be concerned about compliance at facilities in remote locations; however, even an established history of unenforced noncompliance does nothing to protect organizations from adverse consequences. The fact that agencies have failed to detect prior noncompliance or have even acknowledged its existence provides nothing in the way of a legal defense once enforcement is commenced. To be effective at protecting their organizations, environmental professionals need to understand that the longevity of a practice provides little assurance that it complies with environmental laws.

2.2.2.3 Excused Noncompliance

One of the great misunderstandings is that agencies are empowered to excuse noncompliance with environmental laws. While it is true that agencies are generally free to exercise their discretion as to whether to enforce a particular requirement, prosecutorial discretion does not actually constitute permission to break the law.[30] Except in a few cases covered by formal variance provisions, assurances by agencies that a party may disregard an environmental requirement rarely have binding effect. As such, they are free to commence enforcement at any time. Furthermore, the policies and decisions of one agency regarding enforcement do not bind other agencies and private citizens. This means that even if a particular state agency decides not to enforce a particular provision, the federal government or an environmental group remains free to do so.[30,32]

The perception that a facility has been authorized to violate or disregard environmental requirements generally arises from informal oral communications with staff at agencies. For example, an environmental manager at an industrial facility may telephone the solid waste management section at a state agency to ask whether a permit is required for an existing onsite landfill. The person at the agency, overwhelmed by several other projects, tells the environmental manager that the state has extensive regulations associated with industrial landfills but not to worry about them because the state has more important matters to handle. Based on such assurances, the environmental manager may erroneously believe that the facility can continue to operate the landfill. In reality, the agency staff member has no authority to excuse compliance, and the facility is not protected from enforcement commenced by the federal government, the state, or private citizens.

An unfortunate aspect of misunderstanding the extent to which agencies can excuse noncompliance is that it induces some people to consider themselves sophisticated at avoiding environmental requirements. For instance, some environmental managers and consultants seem to delight in the self-perception that they somehow know where the boundary exists between mandatory and optional environmental requirements. Even if these boundaries somehow exist, they will certainly shift in response to the political and personal priorities infused within regulatory agencies; therefore, any belief that one can distinguish between enforced and unenforced violations is largely delusional.

2.2.2.4 Reliance on Technicalities

A propensity to rely on overly narrow interpretations of environmental requirements is a common response to regulatory requirements that impose undue burdens on activities and is fostered by the way many agencies draft and interpret their regulations. Unfortunately, an attitude that reflects the belief that environmental regulations are subject to narrow interpretations leads many parties to serious compliance problems. Although parties are found liable almost every day based on technical constructions of environmental laws by agencies and courts, it is far less common for parties to avoid liability by asserting that regulations should be interpreted narrowly. One reason for this is that courts are reluctant to excuse substantive misconduct merely because an agency could have drafted the intended requirements more clearly.[33] Another reason is that the meaning of some regulations can only be discerned through highly technical interpretations.

Although technical interpretations cannot always be avoided, organizations need to avoid adopting the attitude that expensive or burdensome obligations can be avoided by conjuring up strained rationales regarding the applicability of legal requirements. As discussed in Chapter 5, established judicial doctrines regarding the interpretation of environmental laws will require courts to reject strained interpretations in most situations. This means that relying on implausible interpretations to guide activities can result in violations that are not only serious but also factually easy to prove. There are some things that organizations can do to reduce regulatory burdens, such as participating in the development of regulations, but the use of crafty interpretations to evade regulatory requirements is not one of them.

The temptation to resort to strained interpretations can rear its head when factual information is being considered. For example, the Clean Water Act requires that any person who discharges 100 pounds or more of corrosive hazardous waste in a 24-hour period to navigable waters must report the discharge to federal and state agencies. In such instances, some people are tempted to state that they did not, in fact, "know" about key aspects of the release despite overwhelming circumstantial evidence that establishes the contrary. For example, a person might contend that he is not required to report the spill of caustic rinse water from a 55-gallon drum because he did not "know" whether the quantity exceeded 100 pounds, constituted hazardous waste, or entered navigable waters. In many cases, such facts are actually known or readily ascertainable, and the alleged lack of knowledge is due to a willful decision to avoid confronting the facts. Such temptations should be avoided, as parties rarely succeed in asserting plausible defenses in such cases. Moreover, when this attitude is used to justify serious misconduct such as described in the following case study, the enforcement consequences can be severe.

Case Study: *United States v. Catucci*, 55 F.3d 15 (1st Cir. 1995)

The Facts

Giacomo Catucci owned a building in Providence, RI, where five old transformers were stored. Two of the transformers contained dielectric fluid with high concentrations of PCBs. In 1987, Catucci learned that to dispose lawfully of these transformers would cost at least $16,000. He also knew from an environmental audit in 1991 that the removal and disposal of PCB transformers were strictly regulated. In late 1991, Catucci allowed two scrap-metal contractors to remove a conveyor system from his building in exchange for the scrap value. While dismantling the conveyor, the contractors noticed the transformers and asked the building superintendent if they could have them. Catucci told the superintendent that the contractors could have the transformers if they wanted them but did not ask what they intended to do with them. The contractors, who at that time did not know that the transformers contained PCBs or were subject to special disposal requirements, arranged for a friend with a flatbed truck to help them move the transformers from the property. Afterwards, the three men drained some of the contents onto public highways and the rest at a sand and gravel pit. They then removed the copper from the transformers to sell as scrap and left the carcasses at the sand and gravel pit.

Months later, the government learned of the dumping when the owner of the truck was arrested on unrelated charges. Following the execution of a search warrant, Catucci called the local police to report that the transformers had been stolen. Sometime later, Catucci told investigators that he may have authorized the contractors to remove the transformers but only if they disposed of them legally. At his trial, Catucci testified that he had not given permission for the contractors to scrap the transformers and claimed that he did not know the transformers were missing until the search warrant was executed. Nonetheless, the jury convicted him of unlawfully disposing of PCBs and for failing to report the release of a hazardous substance. The trial court sentenced Catucci to serve 27 months in prison and ordered him to pay for cleaning up the PCBs. Catucci appealed his conviction to the U.S. Court of Appeals for the First Circuit.

Discussion

The First Circuit affirmed the conviction of Catucci, holding that he knew what it would cost to dispose of the transformers lawfully and that such costs could not be recouped by the scrap value that would be received by the contractors. In addition, the court held that the jury was entitled to infer guilt from Catucci's subsequent misrepresentations that the transformers had been stolen. The fact that Catucci did not know the method, time, or place of disposal did not absolve him from responsibility for the disposal and the subsequent failure to report the release.

By relying on a technicality that was patently inconsistent with the requirements governing the disposal of PCBs, Catucci was tempted into committing a substantive offense and suffered the consequences.

2.2.3 ATTITUDES BASED ON PERSONAL TRAITS

Ensuring compliance requires attentiveness, diligence, and sensitivity to compliance requirements. Although many people favor protection of the environment and support the principle of environmental compliance, not all of them have the personal qualities needed to achieve compliance. In order to ensure that the proper people are assigned to positions that entail environmental responsibilities, it is necessary to understand how the personal characteristics of employees can affect the ability of an organization to maintain compliance.

2.2.3.1 Complacency

Complacency is a common source of violations by organizations that are subject to a plethora of requirements. Typically, those responsible for such violations do not necessarily want to violate laws but nonetheless do so because they are complacent about compliance issues. A common characteristic of employees who are complacent about compliance is that they assume that issues that are unimportant to them will be unimportant to others. Another form of complacency justifies incurring violations on the grounds that a low probability of discovery warrants devoting attention and funds elsewhere.

The major problem with complacency is that it allows violations to occur and escalate. This approach to compliance issues is particularly dangerous when the person responsible for a violation misjudges the importance that an agency attaches to a particular requirement. Unfortunately for organizations, it is common for employees at regulatory agencies to have stronger beliefs about the necessity for environmental protection than their counterparts in regulated industries. They also tend to be less familiar with industrial settings and more likely to perceive that industrial activities are harmful to the environment. The result is that agency staff are far less likely to be complacent about compliance issues than may be assumed by people working outside the government.

Complacency can be difficult to counteract, particularly in organizations that have not been subjected to significant enforcement activity or operate in areas where state agencies have traditionally exercised high levels of restraint in enforcement. In some instances, complacent attitudes can be changed by making employees understand that failing to comply with requirements poses long-term risks that can end up being very costly. Common issues that need to be stressed in organizations permeated with complacent attitudes include the need to comply with all requirements regardless of their direct effect on environmental quality and the fact that informal government assurances provide little or no legal protection. Persons who must deal with complacency within their organizations are also advised that persistence can be essential to correcting compliance problems.

2.2.3.2 Sloppiness

Another common problem is the attitude that the incomplete or late satisfaction of environmental requirements will be sufficient to protect organizations from enforcement. It is not unusual for employees to be reasonably familiar with legal requirements yet fall short of what is required. For example, an employee charged with the task of completing the annual document logs for PCB-containing equipment may address the substantive requirements adequately but fail to satisfy seemingly minor requirements such as including the address of the facility in the log. Although minor instances of noncompliance do not always trigger enforcement actions, they frequently are incorporated into enforcement actions motivated by other violations and can contribute substantially to penalty assessments. Sometimes a sloppy attitude toward environmental issues can expose the employer to significant liabilities.[34] For example, a consultant who avoids the effort associated with obtaining sound information and instead uses unjustified assumptions to prepare estimates of air emissions can cause the client to make incorrect determinations about whether a source is subject to permitting requirements.

Sloppy performance can sometimes be improved by making it clear to the employee that a higher standard of performance is required. The performance of some employees improves after oversight is increased, although some employees become even sloppier because they rationalize that someone else will catch their mistakes. Organizations must exercise their own judgment with respect to how much oversight that want to provide to ensure compliance; however, oversight is of limited utility when applied to employees who already have supervisory or high-level technical responsibilities. In general, organizations are advised not to employ people in positions of environmental responsibility whom they do not trust to perform diligently.

2.2.3.3 Disbelief of Legal Requirements

Some people will deliberately commit acts that violate environment laws simply because they cannot bring themselves to believe that a particular legal requirement exists. Because they discount the existence of the requirement, they believe that acts that plainly violate environmental laws are permissible. Generally this attitude manifests itself with regard to matters that involve transient and relatively minor effects on the environment. For example, releasing a single drop of oil into a navigable waterway will produce a small visible sheen that must be reported upon discovery to the National Response Center.[35] However, some people are incredulous about this requirement when applied to small releases, as the sheen generally dissipates before a response action can occur. As a result, they routinely fail to report small releases of oil.

This attitude can also cause people to engage in violations on a larger scale. Examples include requirements under the Clean Air Act and Clean Water Act where refurbishment projects can result in existing facilities being classified as new sources that are subject to enhanced regulatory requirements. Although the regulatory philosophy behind these requirements is that at some point repair projects become reconstruction, some people are unable to accept that maintaining an existing facility can be construed as constructing a new source.

Although the attitude of disbelieving regulatory requirements does not occur with great frequency, it presents significant legal risks. Under the general intent standard that applies to most environmental laws, it is irrelevant whether a person intended to violate the law; therefore, liability can be imposed if the act that violated the law was done knowingly. Furthermore, people who refuse to believe that particular activities are regulated tend to act very openly and thus make their misconduct more likely to be discovered and enforced against.

Addressing the attitude of disbelief can be relatively straightforward when done with some degree of foresight. For example, disseminating spill reporting requirements before such events occur gives people the time they need to ask questions, seek confirmation, and adjust psychologically. Similarly, determining the applicable regulatory requirements early in a project tends to reduce resistance to compliance. Conversely, it can be very difficult to address this attitude on the spur of the moment, as people often need time to consider and accept the existence of requirements that will disrupt their expectations and plans.

2.2.4 Attitudes Involving Interpersonal Relationships

Not only do the attitudes that employees show toward ethics, consequences, and personal performance affect the ability of an organization to achieve compliance, but the attitudes that employees have towards each other can also be of critical importance. Furthermore, efforts at addressing individual attitudes can be completely negated if the managers and employees justify violations by deferring responsibility or misplacing their loyalties. To achieve compliance, organizations must be able to recognize and address employee attitudes that tend to facilitate noncompliance.

2.2.4.1 Fear of Management

An inordinate number of violations occur because employees are reluctant to tell their superiors about compliance problems. The reasons for withholding information vary but typically relate to concerns that reporting a problem will cause some sort of difficulty for the employee. For example, a treatment plant operator may be reluctant to inform management that effluent limits have been exceeded when the underlying cause was his inattention; however, even employees who are not responsible for a violation may decide not to report problems if they fear that management may react negatively. Examples of negative reactions include falsely accusing the employee of wrongdoing or directing feelings of frustration or anger at the employee. In the worst case, employees may fear being branded as troublemakers for calling attention to environmental problems. Employees may also feel that by reporting problems, management may act inappropriately towards the responsible employees who, in many instances, may be friends or close coworkers of the employee.

Fear of management can lead to serious compliance problems, particularly when a prompt action is needed to ensure compliance. For example, several statutes require that environmental releases be reported immediately upon discovery, and any hesitance on the part of employees to report problems can lead to a violation. The enforcement ramifications can be particularly severe when employees document violations in operating logs and similar documents but fail to inform management directly. In such cases, there will be documentary evidence that the organization knew of the violation but failed to address it. To complicate matters, organizations cannot absolve themselves of liability by claiming that the senior managers were unaware of the violations and may damage their credibility if they try to do so.

Managers responsible for noncompliant activities may find themselves in a difficult predicament when employees fail to report possible noncompliance, particularly if the underlying misconduct has criminal overtones. Enforcement staff are likely to be skeptical of claims that the manager did not know of the problems, and the subordinate employees in such situations tend to be uncooperative in establishing a defense.

To avoid such situations, organizations should make their managers understand that it is almost always in everyone's interest to be informed in a timely manner when problems occur and that managers should not discourage impediments to communication through intentional or unintentional acts. Some organizations have addressed this problem by requiring all employees to report compliance issues promptly. This not only makes employees understand that management wants such information but also provides a basis for disciplining employees who withhold this important information. Other organizations have approached this problem by training managers to become better at listening to employees and establishing a policy of respectful communication.

2.2.4.2 Deferral

Deferral occurs when people cooperate in the commission of a violation because they psychologically place the moral burden onto another person. It is rooted in the misguided attitude that moral responsibility can be shifted to others and that personal accountability can be avoided so long as the misconduct was committed or ordered by others. It is a common element in cases where the occurrence of a violation requires the participation of employees at different levels within an organization. For example, subordinate employees may agree to play a role in a violation because they feel absolved of responsibility for misconduct that has been ordered explicitly or implicitly by a manager. Similarly, managers may discount their personal responsibility for even obvious misconduct because they have vested discretion to employees who thereafter chose an unlawful means to accomplish an organizational objective. Although vesting authority in subordinates sounds appealing when cloaked in terms such as "empowerment," it should never be used to excuse management's responsibility to ensure compliance.

An area where large organizations may find themselves particularly at risk is the shifting of accountability among departmental units. Such shifting usually occurs when the lack of organizational guidance allows employees or departments to attribute responsibility for their actions to others. Typically, this pattern occurs in organizations where the departmental units have differing priorities but at the same time depend on each other to accomplish them. An example would be a production department that continues to operate after its pollution control equipment has failed on the grounds that its priority is to make product and the fault lies with the environmental department who failed to keep the equipment operating. Of course, accusations of fault directed toward internal units do nothing to protect the organization from outside liability.

Organizations may also engage in deferral when they allow contractors to commit violations in the generally false belief that they will not be held accountable. For example, organizations will still be held liable for a consultant misrepresenting air emissions in a permit application or a contractor violating regulations governing the removal of asbestos. Organizations need to be particularly vigilant when working with outside contractors, as these parties are not always sensitive to the risks of enforcement and sometimes try to shift the blame to their clients when confronted by agencies.

Deferential attitudes expose organizations to legal risks, as they can be held liable for violations regardless of which employees or contractors committed them. By creating environments where personal accountability has not been communicated explicitly to employees at all levels, organizations increase the likelihood that deferral will occur. This can be particularly troublesome, as deferral often plays a significant role in the most serious violations and can expose several levels of employees to personal liability. As the case study presented below demonstrates, employees and managers that use deferral as an excuse to commit or condone violations are not necessarily comfortable about their participation. Also demonstrated is that reliance on deferral does almost nothing to protect such individuals from liability.

Case Study: *United States v. Sinskey*, 119 F.3d 712 (8th Cir. 1997)

The Facts

The matter, encompassed by the Eighth Circuit's opinion in *United States v. Sinskey*, involved the criminal prosecutions of a corporation and four of its employees. At the heart of the matter was a large meat packing plant in Sioux Falls, SD, that treated some of its wastewater with a company-operated treatment facility and sent the remainder to a municipal treatment plant where it was treated at considerable expense. The plant ran into problems meeting its permit limits for ammonia nitrogen and began to use deceptive monitoring practices to conceal violations.

The plant was continuing these practices in 1990 when the company appointed Timothy Sinskey to be general manager of the plant. Sinskey promoted an operator named Ronald Greenwood to be the treatment plant manager and likewise promoted chemist Barry Milbauer to be the assistant treatment plant manager. Sinskey also promoted Wayne Kumm to the plant engineering position, which made him responsible for duties such as supervising the wastewater treatment plant. The expansion of operations during the early 1990s increased the ammonia nitrogen loadings to the point where it became difficult to show compliance even with the deceptive monitoring practices. The plant responded to this problem by falsifying its monitoring data to conceal the violations and submitting requests to the company headquarters to authorize expansion of the treatment plant. These requests were not approved, and the plant continued its illicit practices.

Milbauer was physically responsible for analyzing the plant effluent to determine the ammonia nitrogen levels. Although he falsely recorded laboratory results, he also maintained a secret log containing the actual measurements. At the trial, Milbauer testified that when he expressed concerns about the falsification, Kumm responded with silence. Milbauer resigned in February 1993, claiming among other things that he was unwilling to continue falsifying the monitoring data. Kumm apparently tried to dissuade him by discussing opportunities for promotion and assuring Milbauer that he had been doing good work. Greenwood testified that he had told Kumm several times that he was using improper monitoring practices "to get them by." He also said that he protested frequently to Kumm about the inadequate treatment capacity. When the expanded operations made it impossible to conceal violations by selective sampling, Greenwood asked Sinskey and Kumm what he should do about the situation but was met with silence. Greenwood interpreted this to mean that he should continue to falsify the data.

Kumm was Greenwood's immediate supervisor and reported directly to Sinskey. Kumm apparently never affirmatively ordered Greenwood and Milbauer to use fraudulent monitoring practices but never ordered them to stop once he became aware that they were occurring. The government maintained at trial that Kumm aided and abetted the criminal violations because he allowed them to continue by failing to object. The government also asserted that Kumm affirmatively encouraged the misconduct by assuring Greenwood and Milbauer that the company would retain lawyers and lobbyists on their behalf if they were ever caught. Greenwood also testified that Kumm had disavowed responsibility on two occasions by stating he was glad that responsibility for the matter was between Greenwood and Sinskey.

Sinskey maintained that he had never been explicitly told about falsified data and thought the plant managed to comply with its permit by sending excess wastewater to the municipal treatment plant. He also asserted that his authorizations of large expenditures to ensure compliance with the permit evidenced that he had no intent to violate the Clean Water Act. However, Milbauer testified that he had told Sinskey that the practice of sampling during periods of low flow was deceptive, and Greenwood testified that Sinskey had been made explicitly aware that the plant was falsifying monitoring data. Sinskey asserted in response that Milbauer and Greenwood had a motive to conceal the violations from him in order to cover up their incompetence and preserve their jobs.

Milbauer, Greenwood, and the company pleaded guilty to violating the Clean Water Act without going to trial. In their plea agreements, Milbauer and Greenwood also agreed to testify against Kumm and Sinskey. The district court convicted Sinskey and Kumm and sentenced them to serve 24 months and 12 months, respectively. Shortly thereafter, Milbauer and Greenwood were each sentenced to 6 months of home confinement. The company paid a criminal fine of $2 million and another $1 million to establish an environmental cleanup fund.

Discussion

Appellate courts generally review issues of the sufficiency of the evidence in the light most favorable to the prevailing party which in this case was the government. With regard to Kumm, the Eighth Circuit commented that the evidence was hardly overwhelming but nonetheless sufficient to sustain his conviction. With regard to Sinskey, the court adopted the reasoning set forth in *Weitzenhoff* that the government did not need to prove that Sinskey knew he was violating the law in order to obtain a conviction.[36] What is notable about this case is that the individual defendants were placed in bad legal positions because of the actions of others. All of them seem to have believed to some degree that actions taken by others would ultimately absolve them of liability. Sinskey inherited responsibility for a plant unable to meet its permit obligations and apparently hoped that his distance from the treatment plant operations would shield him from liability. Kumm was in a difficult position because his intermediate supervisory position gave him substantial responsibility but not to a degree where he felt comfortable ordering that the violations of the permit be reported. Greenwood and Milbauer knew that the monitoring practices were unlawful but apparently felt that the company had given them no feasible alternatives to breaking the law. Had the company been more definite about personal responsibility, it is possible that this sad case could have been avoided.

To minimize the chance that deferral of responsibility will lead to noncompliance, organizations should provide employees and contractors with the explicit directive that everyone is accountable if they play a knowing role in misconduct. To be effective, however, organizations should explore other means of enabling employees to come forward with compliance concerns that may implicate their supervisors or coworkers. A policy of respectful communication can help reduce the possibility of deferral, as can providing a means by which employees can report concerns in confidence to an ombudsman. Establishing formal means by which organizational departments may prevent or resolve conflict over priorities can also reduce deferral within organizations.

2.2.4.3 Vindictiveness

Sometimes organizations employ people whose attitudes are such that their typical response to problems or perceived slights is to inflict punishment or seek revenge. Because compliance matters can give vindictive people great opportunities to punish their employers, subordinates, and coworkers, such people tend to impede the cooperation necessary to address environmental issues effectively and efficiently. Vindictiveness is a dangerous and destructive attitude for several reasons. Employees commonly respond to vindictive managers by concealing problems or postponing their resolution. This has the tendency to prolong compliance problems by making subordinates less likely to inquire whether certain activities violate environmental requirements or to seek corrective action for known problems. In addition, talented employees who must deal with vindictive managers or coworkers often choose to find employment elsewhere. Vindictiveness within organizations can thus result in a loss of valuable talent and makes it more difficult to replace personnel. Vindictiveness within an organization also tends to reduce employee loyalty and the desire to cooperate in solving problems. Because vindictive employees greatly increase the difficulty of achieving compliance, organizations should address this attitude promptly when it becomes evident.

Overcoming vindictiveness is difficult when the underlying cause is insecurity or bitterness on the part of an employee. Sometimes vindictive attitudes can be changed by training or counseling but this usually requires considerable time and effort. In some cases, organizations can reduce the harm caused by vindictive employees by providing subordinates or coworkers with mechanisms by which they can express themselves without fear of retribution. However, approaches that do not resolve the underlying cause of vindictive behavior can undercut the authority of these managers and further increase their insecurity. Considering the enormous risk that vindictive employees create for their organizations, such conduct must not be permitted to continue. If efforts at overcoming vindictiveness do not quickly bear fruit, organizations should strongly consider putting vindictive employees in positions where they cannot impede environmental compliance.

3 The Compliance-Oriented Attitude

Understanding the ramifications of noncompliance and ways that organizations can get into trouble is helpful when trying to ensure compliance; however, achieving total environmental compliance requires environmental professionals to move beyond knowing where the trouble spots lie and to act affirmatively to reach what many perceive to be an elusive and unobtainable objective. Although total environmental compliance requires considerable foresight, concentration, and effort to achieve, it is not impossible. What is required is the proper focus and frame of mind. The compliance-oriented attitude reflects an understanding of what environmental compliance means, what it entails, and the level of attention needed to achieve it.

3.1 WHAT ENVIRONMENTAL COMPLIANCE MEANS

Environmental compliance is obedience to the law and should not be confused with protecting the environment. Although government agencies create environmental laws with the nominal intent to protect the environment, many environmental laws have little direct connection to environmental quality. To achieve compliance, organizations and their employees must realize that they need to obey legal requirements and obey them all the time. The question of whether certain conduct benefits or harms the environment is generally irrelevant to the issue of compliance. While it is good to be concerned about the environment, and such an attitude can help promote compliance, such concerns alone will not shield parties from the adverse consequences of noncompliance.

In fact, attitudes that are strongly motivated by concerns over environmental quality can occasionally be pitfalls when it comes to achieving compliance. For instance, the fact that a particular practice will not harm the environment or that a requirement is too onerous relative to the intended environmental benefit does not justify noncompliance. Thousands of environmental violations have been enforced even though they caused no adverse effects to the environment. Likewise, the fact that there may be scientific or policy reasons for why a requirement is unnecessary or even counterproductive to protecting environmental quality does not justify non-compliance or protect regulated parties from enforcement. Compliance is determined by what the law allows and not by what constitutes the better scientific practice or makes the most sense from the perspective of policy.[37]

Similarly, one does not have to have a positive attitude towards environmental protection to have a compliance-oriented attitude. Although some people equate compliance with protecting the environment, what is really important is the attitude towards obeying the law. This means that a person who feels it is important to obey the law but feels that environmental laws are unreasonable or unduly burdensome may be more suited to achieving compliance than a person with a strong moral commitment to the environment but a less reverent attitude towards the law. In other words, the compliance-oriented attitude focuses on how organizations fulfill their legal obligations and does not factor in questions on whether environmental laws really protect the environment.

In addition to understanding that environmental requirements are not optional, the compliance-oriented attitude also incorporates the realization that achieving compliance can be difficult and requires monetary resources and qualified staff. As a general rule, the environmental field is highly technical, and many compliance functions cannot be adequately performed by persons who lack formal education in the appropriate disciplines. Organizations that show the proper attitude do not assign employees to tasks for which they are not qualified; instead, qualified personnel are hired or assigned or the task is contracted to a consultant. Also, compliance with environmental requirements can be very expensive or extremely burdensome. Some of the ways these burdens can be reduced are discussed in the chapters that follow, but organizations should never use expense or burden as a justification to disregard compliance.

Another important aspect of the compliance-oriented attitude is that it does not factor the intensity of enforcement activity into compliance decisions. Organizations that have the proper attitude towards environmental laws will obey them regardless of whether they are perceived as likely to be enforced. This approach not only reflects good citizenship but is pragmatic, as well. First, it is difficult for organizations to be fully aware of the full scope of government and citizen enforcement, as many enforcements are not well publicized. Second, new avenues of environmental enforcement are continually being explored, and agencies and environmental groups have a history of developing new theories by which to hold regulated parties liable for violations. Furthermore, the statutes of limitations for environmental violations enable violators to be sued for periods extending to 5 years or more after the misconduct, which means conduct that at one time appears to be tolerated by agencies can be intensely enforced a few years later. As the following case study shows, organization should not assume that the absence of prior enforcement means that consequences will never follow.

Case Study: *United States v. Frezzo Brothers*, 602 F.2d 1123 (3d Cir. 1979); 642 F.2d 59 (3d Cir. 1981)

The Facts

The 1970s were a time when the programs created pursuant to the Clean Water Act were getting started. Large corporations often violated with impunity the relatively few requirements that had been introduced, and major stretches of rivers and coastlines were severely degraded. In this regulatory environment, one might doubt that a family-owned mushroom farm would face a serious risk of enforcement because it occasionally discharged recycled water to a small stream.

After investigators observed several instances in which discharges from the farm entered the small stream, the government filed a six-count indictment against Frezzo Brothers, Inc., and the two owners who operated the farm. The defendants contended that the indictment should have been dismissed because effluent limitations for mushroom farming had not been promulgated and also that agricultural discharges did not require permitting at that time. The trial court rejected these defenses on the grounds that effluent limitations were not required before one could be prosecuted for discharges without a permit and that the flow from the composting operation was more like a manufacturing flow than an agricultural one. The defendants also contended that the government must file a civil action or give written notice of a violation before proceeding with criminal enforcement.

Discussion

Despite the fact that this was one of the first criminal cases brought under the Clean Water Act and that noncompliance with the Act was widespread at the time, the government was not obligated to put the Frezzo defendants on notice that they could be subject to criminal enforcement. Likewise, the government was not required to consider the uncertainty of the law when considering whether to commence a prosecution. Neither the relatively small scale of the operation nor the lack of precedent or warning that intermittent discharges could be enforced criminally protected the defendants from a severe application of the government's enforcement powers. As a result, Frezzo Brothers, Inc., was fined \$50,000 and the two individual defendants were sentenced to 30 days of incarceration and fined \$50,000 in the aggregate.

3.2 ASSUMING RESPONSIBILITY FOR ENVIRONMENTAL COMPLIANCE

Environmental compliance does not happen on its own or even when organizations rely on apparently knowledgeable outsiders. To achieve total environmental compliance, organizations must adopt the attitude that they alone are responsible for ensuring compliance. Although advice from agencies, consultants, and attorneys can be helpful in achieving compliance, outside parties rarely understand an organization's activities to the extent needed to ensure full protection from liability. Furthermore, outside parties are rarely exposed to the direct consequences of a violation that results from bad advice and may not be sufficiently cautious when advising clients regarding compliance issues.

A major pitfall that leads many organizations into noncompliance is unwarranted reliance on government agencies for information about their compliance obligations. Although government outreach programs which seek to inform the regulated community about compliance issues are beneficial, many regulated parties attribute too much validity to assurances from agency staff that specific practices comply with legal requirements or that no enforcement will result from a particular violation. Although it is common for regulated parties to consult with government agencies on issues such as how to comply with regulations, interpret permit conditions, and obtain temporary relief from compliance requirements, the information obtained through such communications must be relied on at the regulated party's own risk. Unfortunately, many people believe that if they follow the advice or act on the approval from a regulator then they are protected from the possibility of enforcement.

There are some reasons why reliance on agency advice makes intuitive sense, although reliance on intuition or common sense when ascertaining environmental requirements is fraught with peril. People often feel justified in relying on agency advice because agencies are expected to know what their rules mean, and it seems fair to require agencies to stand behind their word. In reality, except in a few circumstances, government agencies are not bound by the representations made by their staffs. In addition, even agency staff are sometimes uncertain or confused when it comes to applying the legal principles set forth in their regulations to specific situations. Furthermore, neither government agencies nor citizen groups are completely bound by the representations or actions of other agencies. This means that oral permissions and letters may not protect an organization from enforcement should other staff or agencies disagree with the interpretation or suggested course of action.[32,34]

Organizations with compliance-oriented attitudes will not rely on government agencies as their major source for discerning or interpreting compliance requirements, because in many cases agencies are not reliable sources for such information. Although most staff at agencies tend to describe environmental requirements objectively, a considerable number freely exercise regulatory or enforcement discretion that they do not have. For example, there have been instances where agency employees have told parties that they need not obtain coverage under a stormwater permit because their industrial activities have minimal effect on stormwater quality. In reality, these facilities legally must be covered by permits even though their obligations under such permits would generally be minimal. However, by relying on the agency advice, the operators of such facilities expose themselves to enforcement liability.

Another circumstance where agencies tend to disregard their regulations is applying them to matters where the potential for environmental harm appears low and regulatory resources are strained. For instance, an agency may tell a party that they need not apply for a solid waste permit for an onsite landfill because the staff at the office handling solid waste matters does not have the resources to handle additional matters. Similarly, some state agencies have established reputations for ignoring lesser violations because they do not have the enforcement resources to address them. Even when such cases suggest that the possibility of state enforcement is slight, organizations with compliance-oriented attitudes will nonetheless comply with the legal obligations. One reason this approach makes sense from a pragmatic perspective is that other agencies or private parties may step in when an agency fails to enforce its regulations diligently. For example, the U.S. Environmental Protection Agency (EPA) and several environmental groups have a history of filing enforcement actions in states that have been found to be lax in this area. Another reason for compliance in the face of lax enforcement is that agencies often change their attitudes. Such changes are often due to personnel changes, political pressures, or better funding. States that change their minds about enforcing requirements present a difficult situation for regulated parties, as they often have good knowledge of where violations may be found. Organizations that complain that the state agencies "authorized" or at least acquiesced to the violations are sometimes told that the state is not bound by such assurances; furthermore, they more often find that the responsible staff disavow any authorization or acquiescence.

Organizations with compliance-oriented attitudes will also be attentive to areas that are particularly sensitive in terms of enforcement. One of the most sensitive areas in environmental regulation is communications. Communications are sensitive because they can constitute evidence in enforcement proceedings and also because making false statements to the government can be a basis for criminal liability. Anyone working in an environmental field knows that communications that touch on compliance issues are widespread. Organizations communicate internally about compliance issues and often create records such as notes of meetings and telephone calls. Regulated parties may also communicate orally with agencies during inspections and meetings and typically provide copious written communications, such as permit applications, emission fees reports, responses to information requests, and monitoring reports. Improper communication practices can lead to heightened regulatory scrutiny, such as when employees make derogatory comments about compliance motives or exaggerate descriptions of potential requirements. Inaccurate facts in communications to agencies can result in liability for making false statements. Organizations that have the proper attitude towards compliance will devote special attention to developing good communication practices.

One way organizations can implement a compliance-oriented attitude is to provide their employees with guidelines concerning communications and enforce them. In general, organizations that take their environmental obligations seriously manage their documents well. Good document management allows organizations to access information easily and encourages accuracy, completeness, and consistency in communications involving complex legal and technical issues. It also facilitates compliance with records management obligations. For instance, it can be a criminal offense to destroy or conceal evidence that is connected to a judicial or administrative proceedings. Organizations that are subject to litigation or required to respond to information requests will protect their employees by letting them know what documents need to be preserved.

Experience in environmental matters also shows that employees throughout organizations need specific guidance about how information should be communicated, although few organizations provide more than rudimentary informal guidance. For instance, employees should be advised as to where information should be distributed to ensure that key personnel are not deprived of information and also to avoid deluging others with unnecessary information. Similarly, employees should be instructed to be meticulous when using descriptive terms and to avoid inappropriate statements about coworkers in compliance-related documents. This does not mean that communications dealing with negative material should be prohibited but rather that organizations should provide guidance regarding which means and channels are proper for expressing legitimate concerns.

Another characteristic of the compliance-oriented attitude is the willingness to confront problems early. Although hindsight generally shows that problems are most easily solved when acknowledged and addressed early, many organizations tend to defer action for as long as possible. Sometimes this happens because employees are overwhelmed by the demands of the moment and do not have time to address contingent matters. Other times, problems are allowed to escalate because the organization lacked foresight or was not decisive about solving the problem. For instance, some managers cannot bring themselves to act when faced by difficult problems and tend to defer making decisions in the hope that someone else will deal with the problem or that it will somehow go away. In any case, reluctance or inability to confront problems often reduces the options available for resolution and increases the long-term costs.

Organizations that deal with problems early tend to reduce their costs in the long run, even when solutions require short-term effort and expenditures that they would otherwise prefer to avoid. By resolving problems when they arise, organizations can reduce the chance that the scope of the problem will increase, and they retain more control over the outcome. For example, many enforcement issues arise out of disputes that do not directly involve compliance issues, such as when a neighbor's unsatisfied complaint about noise from an industrial facility leads to a call to a regulatory agency which later discovers other issues that must be addressed. In such cases, organizations are better off dealing substantively with the complaint rather than waiting to see if the neighbor complains to the government or the media. The resolution of even apparently frivolous complaints can lead to large expenditures of time and money if they are ignored or dismissed when first presented. Patience and sensitivity can pay off in other ways when potentially contentious issues are resolved early. For example, spending ten minutes explaining to an hourly employee how and why a permit allows certain conduct may seem inappropriate when a manager is overwhelmed with pressing obligations. However, the time and effort it takes to address such concerns are *de minimis* when compared to what may be expended in responding to an agency investigation commenced in response to an employee complaint to the government.

Another practical benefit to dealing early with issues that involve the potential for disputes is that it is often easier to find simple solutions. One reason is that emotional issues are a major reason why most disputes escalate. By addressing substantive issues early, people are less likely to feel that they are being dismissed, ignored, or deceived. Parties whose concerns receive substantive consideration when first presented are more likely to accept reasonable efforts to address their issues; however, parties that feel slighted or otherwise mistreated are far more likely to act in ways that cause disputes to escalate, as they may develop an emotional need to have the offending party punished. When disputes involve regulated parties, the environmental aspects of their activities can become a prime target for punitive efforts.

Another practice of compliance-oriented organizations is that they pay attention to future regulatory issues. While many organizations wait passively as new requirements are promulgated, the better ones participate in legislative and rulemaking activities to influence how the forthcoming requirements will affect them. For example, they will comment on proposed regulations and provide agencies with information regarding how proposed requirements will affect them and how the requirements can be expressed more clearly or made less onerous. Similarly, such organizations plan ahead to account for the time and expenditures that compliance with forthcoming requirements will require. This planning not only facilitates timely compliance when new requirements come into effect, but it also makes organizations more effective at commenting on proposed regulations, as they will have a better understanding of the ramifications associated with the proposal.

4 Foundations for Environmental Compliance

Good intentions by themselves are insufficient to sustain environmental compliance. To achieve total environmental compliance, organizations must prepare a strong foundation from which they can implement the practices critical to ascertaining and following environmental laws. Because environmental issues typically affect several aspects of an organization's operations, committing resources haphazardly to environmental management efforts is neither effective nor efficient. To meet the challenge of achieving environmental compliance, organizations must exercise leadership, adequately staff the environmental function, and establish a compliance program. Organizations that build sound foundations for achieving compliance can address their environmental obligations better than those who add resources as each new crisis reveals another flaw in their approach to compliance issues. Absent a strong foundation, environmental functions tend to deteriorate into sporadic and uncoordinated efforts that inevitably fail to fulfill some compliance obligations.

It should be understood that preparing the proper foundation for achieving environmental compliance requires an organization to deliberate on the issues that affect how well it will be able to execute compliance functions. Leadership is a critical element of the foundation, as employees will need to know what decisions are expected of them in the face of environmental matters. As part of the process of preparing the foundation, senior management should reflect on how they wish the organization's environmental policy to reflect its attitude towards environmental issues. In addition to expressing its commitment to compliance, management needs to ensure that it has sufficient employee talent to meet the technical demands associated with compliance issues. Likewise, the foundation for compliance should incorporate a compliance program through which management's direction with respect to environmental issues is communicated and conformance is monitored.

Care should be taken not to confuse the foundation for achieving environmental compliance with the establishment of an environmental management system. Although systems approaches to environmental management have received a lot of attention, they can be effective only if organizations address the foundational issues affecting compliance. Some specifications for environmental management systems do address foundational issues, but in general they focus on the implementation of procedures and documentation. Although environmental management systems can be useful tools in achieving compliance, they cannot replace the foundational requirements of competence, resources, and commitment.

4.1 THE ELEMENT OF LEADERSHIP

The technical and legal complexities associated with environmental matters demand that senior management thoughtfully address how the organization will address compliance issues. In some organizations, the tendency is for management to want to adopt positions that strongly favor the protection of environmental quality. In others, management may wish to avoid any effort at adopting a formal position, as environmental issues play no significant role in the objectives they consider important. At their extremes, attitudes of senior management towards environmental issues can be counterproductive to achieving environmental compliance, as they may reflect unrealistic expectations of either the difficulty of achieving compliance or the consequences associated with not meeting environmental requirements. For senior management to succeed in leading their organizations to total environmental compliance, they must be committed to achieving a demanding objective. It is unrealistic to expect that most senior managers will be expert at environmental matters, but they should understand that environmental matters are technically complex and that compliance will not happen on its own accord. Too many senior managers make the unrealistic assumption that if they express their commitment to environmental compliance, the rest of the organization will follow. Unfortunately, this is rarely the case.

Although many managers claim respect for the environment and promise that their organizations will behave responsibly, they are not always credible. First, credibility with respect to environmental compliance is something to be established, as people have a natural tendency to discount assurances that management is willing to invest in activities that produce no obvious financial return. Second, credibility is easily eroded by actions attributable to management that do not reflect their professed doctrine. To have any hope of credibility, management not only must express its commitment to environmental compliance but must also ensure that the organization conforms as a whole to that commitment. Credibility with regard to environmental commitment is a fragile thing that will disintegrate if management action flagrantly conflicts with management representations. Credibility will also deteriorate over time if the organization fails to abide by management directives regarding compliance. Actions and decisions that can be construed as tolerance of noncompliance, especially for pragmatic reasons, will eventually undercut efforts to impose a high standard of environmental performance, especially when senior management has no direct participation in ordering noncompliant activity but nonetheless appears to acquiese.

To build a solid foundation for environmental compliance, senior management must show purposeful leadership when it comes to compliance matters. Success is dependent on senior management having respect for the law, as they are the ultimate decision-makers and serve as models for the rest of the organization. Furthermore, management must understand that respect for environmental laws is an issue of citizenship and does not require that they become environmentalists. In fact, ethical companies may oppose legislation or challenge the promulgation of regulations which they believe are ill-considered, so long as they do not knowingly disregard them.

Although a compliance program should be based on citizenship, this does not mean that organizations need be insensitive to issues that affect the environment. Many environmental matters do not involve compliance issues, and organizations are free to exercise their conscience with respect to their actions. Business considerations may reflect whether and how an organization participates in environmental cleanups or acquires contaminated properties. Similarly, organizations may want to consider community issues when it comes to decisions such as whether to modify their operations to reduce adverse effects on the environment. These kinds of matters are worthy of management attention but should not be confused with environmental compliance issues. Confusing environmental issues with compliance issues can lead to the paradoxical situation where environmentally conscious organizations expose themselves to serious enforcement liabilities. For instance, an undue emphasis on environmental issues can lead to an allocation of resources that places environmental quality above environmental compliance. However, achieving total environmental compliance does not depend so much on sensitivity to the environment as much as sensitivity to the obligations established by law.

4.1.1 Defining Environmental Policy

When considering the issues associated with developing organizational policies with regard to the environment, it is useful to understand how attitudes and perceptions of regulated entities have evolved and why organizations typically respond to environmental regulations differently than they did when the modern environmental movement began. Activities affecting the environment became heavily regulated during the early 1970s, and the predominant reaction by businesses and municipalities was to resist additional regulation. In large part, this reaction was due to a reluctance to internalize environmental costs when the prior experience was that wastes could be released to the environment at nominal cost. Environmental expenditures were viewed negatively by many managers who believed that the economic benefits of unimpaired industrial activity far outweighed the costs of pollution. New environmental laws were not always taken seriously, particularly when enforcement was lax and legal and legislative challenges were grounds for optimism that environmental requirements would be made less stringent. Even when requirements were seriously considered, challenges to new regulations were common, and pollution control equipment was added begrudgingly. In addition, organizations often staffed facilities with environmental personnel sparingly and belatedly.

The regulated community began to change its attitude towards environmental issues during the 1980s. One reason for the change was that the high but unavoidable costs associated with contaminated properties caused many managers to view environmental issues as realities that needed to be addressed as legitimate business matters. This willingness to confront environmental matters also encouraged managers to be more realistic when allocating resources to environmental issues. In addition, regulated parties began to shift away from blindly resisting new regulations and became more active in attempting to influence how regulations were developed. As regulated parties became less combative, many changed from a strategy of paying millions of dollars in the hopes that legal efforts could avoid liabilities and instead focused on increasing the cost-effectiveness of their environmental efforts.

Another aspect of the 1980s is that the ease and low cost of addressing some kinds of environmental problems led many organizations to believe that most investments in environmental projects could have positive returns. For example, some companies were able to report that they had substantially reduced their releases of toxic emissions merely by developing more accurate but less conservative estimates. Many organizations found economically viable substitutes for products that generated hazardous wastes. Such experiences either cost very little or resulted in reduced costs; however, the resulting publicity over the success of pollution prevention efforts probably over-represented the economic benefits. Regardless of the actual benefits, the perception that environmental regulation did not necessarily restrain economic competitiveness caused many organizations to stop viewing environmental expenditures as something that should always be minimized.

Another important development during the 1980s, and perhaps the most important concerning environmental compliance, was that the government and citizen groups began enforcing the literal provisions of environmental laws with great success. Organizations that were found to have committed technical violations ended up paying large penalties and entering into binding commitments to bolster their environmental performance. The increase in enforcement activity not only increased the risk of penalties but also reduced the uncertainty over whether environmental compliance would put organizations at an economic disadvantage with respect to their noncompliant competitors. Given the increased certainty of enforcement and the risk of high penalties, many managers began to view compliance efforts as an essential element of organizational stewardship. The increase in criminal prosecution of individuals when deliberate noncompliance could be proven also caused many managers to shy away from deceptive practices intended to conceal noncompliance.

Organizational attitudes toward environmental regulations that have developed since the 1980s have moderated. For example, the optimism that environmental projects would usually pay for themselves has diminished as companies have found that the opportunities for economic pollution prevention are far less extensive than was initially represented by trade publications and agencies. Likewise, many companies have found that consumers are not receptive enough to environmentally conscious products to allow for recovery of additional costs through higher prices or increased market share. However, organizations are far less likely than they were in the 1970s to assume that environmental projects will adversely affect profitability. With respect to enforcement, organizations have found that the risk of enforcement remains high and that compliance remains important. The overall effect of the changes in attitudes is that organizations are becoming more pragmatic about environmental issues. Rather than viewing environmental issues as mostly "win-win," many organizations are seeing them as neither good nor bad but rather as business issues that need to be managed in a way that results in the best long-term bottom line.

Because environmental enforcement and collateral problems present genuine risks for noncompliant organizations, management should work less on developing aspirational goals and concentrate instead on developing pragmatic policies that establish a firm standard for organizational conduct. One reason why pragmatic approaches are favored for policies is that conformance to overly optimistic principles tends to deteriorate as people question the validity of the assumptions that underlie such principles. For example, a low-level manager confronted with a difficult ethical dilemma may feel inclined to dismiss a policy provision espousing lofty generalities because of doubts that the drafters really understood the issues faced in the real world. To be effective as a compliance tool, policies need to set forth a yardstick by which managers can assess the propriety of decisions that affect regulated activities. By defining expectations clearly, management not only provides guidance to employees faced with difficult decisions but also establishes a tangible basis for evaluating employee performance.

Management also needs to be careful that the premises supporting the policy are realistic. For example, a policy that states that the organization will demonstrate good citizenship by complying with legal requirements will likely be more effective than a policy that is based on ambiguous, erroneous, or debatable assumptions. For example, policies that tell employees to act in ways consistent with fostering the public welfare are ambiguous because the potential for disagreement over what constitutes a benefit to public welfare tends to diminish the clarity of how management expects employees to act. Similarly, policy statements that rely explicitly on premises such as environmental expenditures will lead to increased operating profits will be viewed skeptically by many employees should the premise be discredited.

Organizations often develop their policy towards compliance in conjunction with setting policy on other environmental issues. Although there is nothing inherently wrong with this approach, they need to be careful not to confuse compliance issues with the less compulsory objectives encompassed by the policy. Because organizations should not risk compromising the principle of total environmental compliance by associating it with more elective goals, they need to be careful how they express issues that do not necessarily involve compliance issues. While principles such as working cooperatively with communities to protect the environment are admirable, organizations need to draft them in ways that enable them to decline acting when the demands are unreasonable. For example, a company may willingly provide local residents with bottled water while it cleans up the groundwater supplies it has contaminated but will probably refuse to build a new headquarters complex for a local group of environmental radicals. By making the policy clear that compliance is always necessary and other environmental matters are subject to the rule of reason, management can avoid the risk of unintentionally diluting its message regarding compliance.

Probably the best way to express management guidance regarding compliance is to draft a clear unadorned statement that the organization will comply with all environmental laws. If desired, organizations can elaborate the policy statement by providing guidance regarding individual duties and how employees are expected to act in specific situations; however, management should be careful not to obfuscate or weaken their policies by incorporating ambiguous text. For instance, many organizations express a policy of going "beyond compliance" without making clear what is meant by the term. Undefined and typically unnecessary qualifiers to a directive to comply with environmental laws can confuse employees and diminish credibility. For example, the term "beyond compliance" could be construed to mean that the organization will treat its wastes to levels below permit limits, implement environmental projects that are not required by law, or agree to follow any request to restrict operations. Furthermore, such statements invite nonconformance with the policy when applied to situations where they would require unreasonable action on the part of the organization. For instance, it is unlikely that a company that barely but consistently meets its permit limits would actually install additional control equipment for the sole purpose of going beyond compliance.

With regard to promoting compliance, policy statements must provide employees with sufficient information to know what is expected of them. Unfortunately, providing the necessary level of detail in a policy statement may conflict with other uses of the policy statement, such as public relations where concise statements are preferred. Furthermore, management may be reluctant to discuss important issues such as employee discipline, hazardous wastes, and reporting responsibilities in the policy statement for fear that they will not make a positive impression on the public. One approach to overcoming this problem is to express the policy in separate but consistent policy statements. The general principles underlying the environmental policy can be set forth in a concise policy statement that is suitable for public communication, and more detailed information for the benefit of employees can be described in a more comprehensive employee policy. Consistency between the statements can be ensured by drafting the employee policy to complement the general policy.

Although general policy statements should limit the discussion of compliance issues to clear expressions that the organization will comply with all environmental laws, policy statements intended for employees should provide more information on the environmental aspects of working for the organization. Specifically, the employee policy statement should describe their responsibilities with regard to compliance, the duty to report concerns to management, and the resources available to ensure compliance. Such statements should be no longer than a few pages and written in plain language.

After an introductory statement explaining its purpose, the employee policy statement should describe the responsibilities of employees with regard to matters affecting environmental compliance. The overall responsibility of all employees to obey all environmental laws should be forcefully and clearly stated. In addition, the policy statement should inform employees that the organization will view most negatively the willful or negligent failure of any employee to comply with a legal requirement and that violations for those reasons will subject the responsible individual to discipline. The employee policy statement spells out that employees are expected to exercise good judgment with respect to environmental matters. Specifically, this part of a policy statement should address common excuses that employees offer to justify noncompliant conduct. For instance, employees can be advised that they need to consider how environmental problems can be minimized or avoided before they start activities and that they need to consult with persons who can provide proper guidance if they are unsure of environmental requirements. Another important function of the good judgment directive is that it emphasizes that employees cannot defer the responsibility for environmental compliance to others.

The duty to communicate is another important issue to address in the employee policy statement. The statement should make reporting of existing or imminent compliance problems a mandatory duty. By requiring employees to report environmental problems immediately to a supervisor or other appropriate person, organizations can reduce passive tolerance of wrongdoing or other forms of noncompliance. In addition, emphasizing the duty to report problems can make it clear to employees that failing to notify a supervisor can result in discipline and reinforces the message that compliance is to be a fundamental practice. It also serves as a deterrent to employees who may be willing to disregard compliance requirements, because they will know that the reporting obligation makes discovery and discipline more likely.

Organizations must be realistic and understand that there may be practical restraints on employees' willingness to report environmental concerns. The most significant of these is the fear of retaliation by supervisors, although coworker pressure can also discourage reporting. It is critical, therefore, for the employee policy to state that employees will not be treated adversely in any way because they reported a compliance concern in good faith. In addition, supervisors and managers should be advised that retaliatory action against employees who express concerns about compliance issues is prohibited conduct and will subject anyone who acts adversely towards such employees to disciplinary action. It is also a good idea to explain what is meant by good faith when reporting a concern. Most reports will be made in good faith. Exceptions to the non-retaliation policy should be limited to attempts at extortion or reports made in a violent or abusive manner.

Organizations should realize that many employees will still fear retaliation even in the face of an explicit policy against adverse treatment. For many organizations, it may be worthwhile to establish a confidential reporting system to further encourage employees to report concerns. These systems work by allowing employees to report problems to an ombudsman who will relay the information to management but preserve the identity of the employee; more information on how to set up a confidential reporting system is provided later in the next chapter.

Because training serves an important role in ensuring environmental compliance, organizations may want to include a requirement in the policy statement that employees must attend all training sessions to which they are assigned. The policy statement may also want to address who has the ultimate responsibility to see that employees are adequately trained. For example, plant managers could be assigned the responsibility to ensure that employees at the facility receive proper training, although in most cases someone other than the manager will actually provide the training.

Although the employee policy statement should address issues that apply to all employees regardless of their level in the organization, it should also emphasize that managers have special duties with regard to environmental compliance. One of these special duties is to know what environmental requirements apply to their operations and ensure that those requirements are consistently met. Likewise, managers have a responsibility to communicate important information to their superiors and, when relevant, to other employees within the organization. The policy should make clear that senior managers and environmental departments are to be kept informed about important environmental developments.

One of the benefits of developing an employee policy statement is that it provides senior management with the opportunity to consider and set policy with respect to significant or difficult compliance situations. For example, many companies have never set forth official positions on whether production facilities should be allowed to operate when pollution control systems fail and permit limits are exceeded. This kind of situation puts operations and plant managers under great pressure when they are uncertain of how senior management expects them to act. By establishing policy for these situations, senior management can provide the guidance the subordinates need to act in accordance with the organizational ethic.

Issues other than production shutdowns can also be addressed as policy matters. Other situations to consider include the organization's policy toward allowing access to facilities by government regulators, permitting responsibilities, and responding to requests for information. Although the directive to obey legal requirements is always the guiding principle, managers generally appreciate specific guidance on how they are expected to handle compliance situations that fall outside the daily routine.

Another aspect of organizational policy that should be addressed is the allocation of responsibility and the resources available to assist in compliance efforts. For instance, organizations can designate which positions have the overall responsibility to ensure compliance at individual operations and facilities. For example, large organizations may want to designate plant managers as the individuals that are responsible for ensuring that compliance requirements are identified and met at their facilities. Although it is unreasonable in many cases to expect plant managers to fulfill every aspect of compliance personally, it is nonetheless imperative that an individual be identified as having responsibility; otherwise, confusion and deferral of responsibility will become more likely. One way to ensure fairness when assigning personal responsibility to individuals who lack the qualifications to perform environmental functions is to allow these individuals to use outside resources to address compliance issues. For example, although a plant manager in a large organization is responsible for seeing that the plant operates in compliance, he or she is free to request assistance from the corporate environmental department; however, the burden of ascertaining potential requirements remains with the plant manager, and the overall responsibility for the compliance function cannot be shifted to the environmental department.

4.1.2 Communicating Direction

Senior management must communicate its expectations of compliance to the rest of the organization if it is to avoid situations where critical ethical decisions are made by subordinates throughout the company with little or no guidance as to whether those decisions conform to organizational goals. Furthermore, because attitudes towards compliance vary substantially throughout society, failing to communicate the expected level of performance makes poor decisions inevitable by forcing subordinates to sort out compliance issues on their own. Another reason for communicating that compliance is required throughout an organization is to fulfill the senior management responsibility to avoid the consequences of noncompliance.

While senior managers must necessarily delegate the responsibility for much of the organizational decision-making to subordinate managers, neither they nor their organizations can abdicate the ultimate responsibility for violations on the grounds that the wrongful acts were initiated by lower level employees; therefore, it is in the self-interest of senior management to communicate to subordinates that environmental compliance is required. In some cases, senior managers have been held criminally liable for illegal acts committed without their active consent.[38] More frequently, managers have been terminated or demoted following a downturn in organizational fortunes caused by enforcement problems. By making their direction known, senior managers can not only prevent compliance problems but also establish a personal reputation for ethical conduct. Such a reputation is good for the organization, and it can protect the individual from adverse consequences during difficult times. Some managers fail to realize this benefit until a crisis occurs, but it is difficult to develop a good reputation once problems arise, and prudent managers will make themselves known as sticklers for compliance long before events subject their ethics to scrutiny.

While communicating the management directives need not necessarily occupy a lot of management time, expressing that compliance is mandatory must be viewed as an ongoing activity. Unless constantly reminded, employees may come to doubt the validity or importance of policies and practices regarding environmental compliance. Changes in personnel also require ongoing education as employees take on new responsibilities. Organizations should incorporate the message that compliance is mandatory in their everyday business practices to integrate it into the corporate culture. There are many areas where this can be done, but two significant ones from the organizational perspective are requests for expenditures and personnel evaluations. For example, most organizations have formal procedures for authorizing capital expenditures in which requests are submitted that contain information about the cost, purpose, and payback period associated with proposed expenditures. Compliance issues can be made an element of the approval process by requiring that managers provide information such as permit requirements and the effect that the proposed project will have on compliance costs. In addition, the persons preparing the request can be required to certify that they have identified and are prepared to address all related compliance issues. Appraisal of employee performance is another area where conformance to the compliance policy should be formally incorporated into business practices. By doing so, organizations can better reward employees who manage compliance issues effectively and discourage employees who may be tempted to ignore or disregard compliance issues.

Employee training programs present good opportunities to teach employees about the organization's compliance policy. In many instances, training on the policy can be incorporated into existing training programs, such as those set up to satisfy the training requirements under hazard communication standards and stormwater permits. Some organizations may want to train managers and hourly employees in different sessions so they can emphasize particular areas. Organizations should also ensure that newly hired employees receive adequate training before they are put in positions where they can cause compliance problems. In addition to formal training, employees can be provided information outside of formal sessions. For instance, posters and handouts can be left in employee areas such as lunchrooms or coffee stations. Similarly, policy information can be included in documents such as operating procedures and financial manuals.

Environmental auditing can also be used to emphasize the need to conform to the compliance policy. While environmental audits are ostensibly addressed toward evaluating whether a business activity is conforming to policies and procedures, they also communicate to employees that management is interested in their environmental performance. This expression of management vigilance can enhance the degree of communication between supervisors and line employees about environmental matters and encourage timely training. Although auditing should not be considered a primary tool for incorporating the policy into organizational practices, the oversight function lets managers know that their attentiveness to the policy will be reviewed occasionally.

4.1.3 Allocating Resources

Senior management is ultimately responsible for allocating the resources that need to be devoted to environmental issues. For the most part, resources will need to be allocated in terms of capital, operating expenses, human resources, and outside services such as consulting and legal. Generally speaking, most of the information senior management will need to make allocation decisions will come from subordinate managers. Although most senior managers are experienced at allocating resources, some issues associated with environmental compliance can prove troublesome if managers base their decisions on erroneous assumptions. For example, some managers incorrectly assume that small or uncomplicated facilities will be subject to simpler regulatory requirements than expensive, complex operations. In reality, even small operations can entail complex regulatory issues and lead to serious compliance problems if given insufficient resources.

Another area where senior managers may misunderstand compliance needs when it comes to allocating resources is underestimating the level of employee qualifications that will be needed to manage environmental matters. Many senior managers worked at the production level in eras when much less attention was given to environmental matters. As a result, their experience tells them that environmental matters can be handled effectively by persons with no particular environmental expertise and often as an ancillary responsibility. However, environmental regulation has evolved to the point where the issues at many facilities can be managed effectively only by trained professionals. The risks associated with failing to staff the compliance function with adequate talent should be obvious but are not always considered by senior management. Furthermore, reliance on unqualified staff can impede future efforts to allocate resources, as senior managers will be less likely to receive the information needed to make good decisions if their subordinates are unable to understand compliance issues fully.

Although managers must necessarily rely upon subordinates to determine what environmental projects are necessary, they must be cognizant that expenditures need to be made wisely. In particular, managers should not fall into the trap of believing that any environmental expenditure will promote compliance, as many environmental matters do not involve compliance issues. For example, increased recycling may or may not make sense from a business perspective but such programs typically do not involve compliance issues; therefore, managers need to be able to distinguish projects that involve compliance issues from those that promote environmental quality. Making such distinctions does not mean that projects that promote environmental quality are not worthy of serious consideration, but if the goal of the organization is total environmental compliance, then projects that pertain to compliance must be given higher priority.

Decisions to cut or delay environmental expenditures are unwise when they weaken the organization's ability to ensure compliance. Managers should review proposed environmental expenditures with regard to their effect on ensuring compliance. Some expenditures that are greater than necessary to meet the minimum compliance requirements may nonetheless produce a valuable return if they reduce the potential for serious regulatory difficulties. For example, by installing monitoring equipment not required by regulations or permits, parties can collect data that can facilitate compliance certifications and future permitting efforts. Such equipment will easily pay for itself when used to reduce compliance burdens or to facilitate the issuance of permits.

Finally, managers should be aware that their actions with respect to approving environmental expenditures may be subject to special scrutiny in enforcement proceedings. It is not unusual for plaintiffs in enforcement actions to take testimony from the defendant's officers to establish that the organization deliberately avoided compliance expenditures. In extreme cases, corporate officials have been criminally charged for failing to approve adequate expenditures.[39] Managers that are involved in processes that entail the consideration of whether to spend funds to ensure compliance are well advised to reflect how their decisions will look when subjected to hindsight.

4.2 ESTABLISHING ENVIRONMENTAL COMPETENCE

Leadership and talent are essential components of an environmental compliance program but are often overlooked when organizations seek to improve environmental compliance. One reason for this is that managers do not always understand how environmental laws affect their operations and underestimate the qualifications needed to manage compliance issues. To achieve total environmental compliance, management must see that the environmental function receives the staffing needed to address environmental issues effectively; otherwise, compliance issues will start to demand time that managers need for other duties. By understanding what level of competence they need to ensure compliance, organizations can obtain a better return for their compliance expenditures and reduce the amount of time that its management deals with environmental issues.

There are several indicators that demonstrate that the environmental management function of an organization is not being adequately supported. The most obvious is that the organization becomes the subject of an inordinate amount of enforcement activity. Incurring legal expenses and penalties generally makes a loud statement, regardless of the accounting system used, and often results in more resources being allocated to environmental matters. However, organizations that reach this stage typically find that achieving compliance is far more expensive than what would have been the case had the initial staffing and expenditures been adequate. Furthermore, unless an organization understands what qualifications are needed to deal with compliance problems successfully, spending more to bolster the environmental staff may fail to achieve the desired objective of ensuring compliance.

A less obvious symptom of inadequate staffing is that environmental matters begin to consume an inordinate amount of senior management time. In severe cases, managers at companies with inadequate environmental staffs can find themselves spending up to 80% of their time attending meetings with agencies or otherwise dealing with environmental issues. Although the willingness to devote time to environmental issues is commendable, senior managers are rarely expert at such matters and the long-term diversion of management time can adversely affect the organization's performance in other areas. Organizations that hire qualified people to handle most of the environmental work also allow their senior managers to concentrate on their core responsibilities. This usually results in enhanced productivity and better compliance.

Another indication that the environmental function is inadequately staffed is that the organization shows a tendency to perform its environmental duties sloppily. Sometimes sloppy work results from staffing at levels that do not allow employees sufficient time to execute projects properly. Other times, the cause of sloppy performance is that the employees with environmental duties lack the interest or qualifications to do good work. Insufficient time or competence can lead to situations where senior managers must step in to devote their time to resolving compliance problems. For example, an inadequately completed application for a permit can lead to permit violations, which then lead to high-level meetings to resolve enforcement actions.

The time and cost expended to resolve such matters is grossly disproportionate to the time required to complete an application properly. Organizations that invest in the highest level of environmental competence do much to reduce long-term unproductive expenditures.

4.2.1 The Senior Environmental Officer

The skillful management of environmental issues can give organizations a competitive advantage by dealing with matters in a way that reduces their long-term costs; however, organizations are rarely able to manage such issues skillfully without the oversight of a competent senior environmental officer. Among the skills such officers provide their organizations are communicating special insight to the board and senior management, allocating resources to where they are most productive, and paying sufficient and timely attention to regulatory developments.

Selection of the senior environmental officer can be one of the most important management decisions at a company. In addition to compliance matters, many organizations face environmental issues such as cleanup liability, public relations, and social pressures to reduce the environmental effects of their activities. These issues can be legally and technically complex and often entail internal and external conflict. Furthermore, missteps by the senior environmental officer can subject organizations to costly encumbrances. Considering the daunting challenges that this leadership position entails, organizations should seek only the most qualified individuals when filling this position.

Many organizations have made the effort to hire highly skilled individuals with stellar credentials to be their senior environmental officers. In general, these organizations should be commended for recognizing the difficulty of performing this function well and setting commensurate standards for filling the position. Typically, the companies with the best senior environmental officers look for people who are skilled at coping with both internal and external conflict and able to analyze difficult technical and legal issues. Another trait that is highly sought is the ability to pay attention to the details of ongoing matters while at the same time showing foresight with respect to future issues. Unfortunately, not all organizations appreciate the need to retain senior environmental officers of the same caliber they require of their other senior officers. The result is that some organizations effectively establish a lower standard for this position relative to what they require for the executives responsible for managing production, finance, and sales activities. For the most part, these organizations act this way because top-level managers do not comprehend the difficult demands of the position and are unable to recognize substandard management of environmental matters.

There is a general tendency on the part of organizations who have senior environmental officers to favor candidates with either political or technical credentials. Those organizations that favor political credentials tend to look for someone who has held an appointed position at a regulatory agency and knows how such agencies function. Organizations that favor technical credentials look for people with technical backgrounds and managerial experience. One reason why organizations tend to favor one type over the other is that relatively few people have backgrounds that combine strong political and technical experience. There are advantages and disadvantages associated with either of these backgrounds, and organizations should understand them before hiring a person to be their senior environmental officer.

One of the major strengths of senior environmental officers with political backgrounds is that they generally have strong credibility with regulatory staff. This is particularly useful for organizations that are involved in controversial activities or need to improve their relationships with regulators. Another strength is sensitivity to how regulatory agencies react to environmental matters, as this is a useful perspective when organizations make decisions that have environmental ramifications. The major potential weakness associated with a candidate having a political background is a lower level of technical skills.

Although persons with technical backgrounds do not always have substantial experience with enforcement matters and may not be as sensitive to regulatory attitudes as a person with a political background, they often serve well as senior environmental officers for organizations that face substantial capital and operating expenses because of environmental requirements. Another strength is that persons who have succeeded in technical areas tend to be careful in their work habits and generally demand the same of their subordinates. These high standards can greatly enhance the overall environmental performance of the organization's environmental function.

Most organizations need senior environmental officers whose skills enable them to handle competently both the political and technical aspects of environmental compliance. When selecting a person for this position, organizations should evaluate how well a candidate will be able to execute the various functions of the position. Regardless of the senior environmental officer's background, the person needs to have excellent interpersonal skills and ethical credibility. Interpersonal skills are important because the senior environmental officer can be expected to communicate in many settings. Not only will the senior environmental officer be required to deal with other senior management, but he or she will also likely have to face government agencies, the media, and communities. An inability to interact effectively and appropriately with any of these parties can significantly handicap an organization's ability to manage environmental issues.

The senior environmental officer's attitude towards ethical conduct will over time establish a firm imprint on the environmental performance of the organization. If this person expects high-quality work and the exercise of foresight in subordinates, the organization can expect to improve its compliance situation over time. Tolerance of noncompliance by this person will be reflected by a lowered standard of performance throughout the organization. Similarly, attitudes about whether to interpret legal requirements aggressively or conservatively will evidence themselves in the organization's vulnerability to enforcement. Organizations that employ senior environmental officers with previous responsibility for serious noncompliance may face additional challenges in matters involving regulatory agencies because of the mistrust or skepticism such a history will generate.

One of the critical functions of the senior environmental officer is to advise the Board of Directors and other senior managers on important environmental matters. To perform this function competently, the senior environmental officer must be able to determine where environmental resources are needed and know how future regulatory developments will affect compliance. Assisting senior management requires a person with a comprehensive knowledge of environmental matters and a high degree of confidence in their positions. Because respect by peers is important to perform successfully at this level, the senior environmental officer must have demonstrated management skills.

In addition to interacting with other senior managers, the senior environmental officer is responsible for administering the environmental activities of an organization. Among these duties are directing the efforts of environmental employees and ensuring that they are adequately supported. The senior environmental officer is also responsible for seeing that the organization competently monitors and evaluates its environmental performance. The ability of this person to hire and retain qualified environmental staff will have a significant bearing on how well the organization avoids compliance problems. Organizations relying on senior environmental officers who cannot relate well to their employees can expect lapses in compliance because the environmental staff will lack the stability and level of comfort needed to establish sustained compliance.

The senior environmental officer must also be skilled at communicating with government agencies, the media, and the public. Organizations are sometimes called on to have a senior management presence at meetings and events concerning environmental matters, and the senior environmental officer is usually the favored candidate because of the expertise and sensitivity required. For consistent success in these settings, organizations must employ senior environmental officers that have the knowledge, maturity, and professionalism necessary to act appropriately when under pressure.

Another reason to be careful about selecting a senior environmental officer is that organizations and senior managers have no choice but to depend on the person occupying this position to provide the guidance and judgment necessary to avoid penalties and collateral litigation. As discussed in Chapter 2, corporate managers generally cannot avoid personal responsibility by delegating compliance matters to subordinates. With regard to environmental matters, the delegation of the compliance function will largely fall onto the senior environmental officer. Because this delegation of responsibility will not necessarily protect other managers from liability, organizations need to ensure that this position is filled by someone who is truly competent and committed to administering the compliance function.

4.2.2 Environmental Staffing

Although much has been made of the importance of corporate compliance programs and environmental management systems, whether organizations succeed at achieving compliance is ultimately determined by the acts of its employees. Unless employees execute the tasks necessary to ensure compliance competently and in a timely manner, even the best compliance programs and management systems cannot protect organizations from liability. To establish a sound foundation for achieving compliance, organizations should take great care to hire individuals that are qualified for environmental duties and to foster their professional development. Furthermore, organizations should monitor employee performance and act promptly and responsibly when an employee's performance gives cause for concern.

4.2.2.1 Employee Qualifications

The range of skills needed to carry out environmental functions varies according to the regulatory setting in which the organization operates. At one extreme, a dry-goods store may need limited talent if its only environmental obligation is to sort paper from plastic in a manner sufficient to comply with the local recycling ordinance. At the other extreme, a chemical company may need a large staff of professionals capable of performing activities such as process engineering and modeling. While there are no bright-line rules for determining what qualifications are needed to ensure environmental compliance, organizations need to give thoughtful consideration to the level of talent necessary to ensure compliance.

Perhaps the biggest obstacle to hiring qualified environmental employees is that many organizations do not understand how difficult the compliance function can be to execute. One reason for this is that organizations do not always assess what skills are needed to address the technical issues associated with their compliance issues. Organizations can greatly benefit from examining the environmental aspects associated with their activities and determining what skills and knowledge are required to perform the tasks related to compliance. For example, facilities that control air emissions and treat wastewater discharges need employees who can competently operate pollution control equipment. Similarly, organizations that must report releases or estimate emissions need employees who understand the processes sufficiently well to be able to develop accurate estimates. Although there is some flexibility in how to assign employees to environmental duties, organizations that do not identify the skills and knowledge necessary to achieve compliance will not be able to ensure they employ the required talent.

Another reason why organizations sometimes assign unqualified employees to environmental positions is that they do not understand the difference between being able to understand environmental issues and having the technical skills to analyze and solve environmental problems. The growth in demand for environmental professionals has allowed many people to enter the field who lack the credentials that were required when the field was dominated by engineering professionals. Although such employees may be competent at performing certain activities, it is unreasonable to expect them to perform serious design and process work. While such persons may know a lot about their subject areas, this knowledge does not make them competent to perform tasks that require years of intense study and experience to master.

Regardless of technical abilities, environmental employees must also have other skills if the organization is to achieve total environmental compliance. Managing compliance issues requires employees who pay attention to detail, are good at determining priorities, readily comprehend legal and technical concepts, and are skilled at communicating. Such persons are hard to find, which is one reason organizations should strive to retain their best environmental employees. Also, organizations should distinguish between the compliance function and other environmental activities when assigning employees. For instance, employees who are technically skilled but otherwise insensitive to compliance matters may be very effective in handling issues such as environmental cleanups but considerably less suited to positions that involve conformance to regulatory requirements.

Finding qualified employees is not always easy. Faced with the realities associated with the number of qualified professionals available and their salary requirements, many organizations have been forced to staff some environmental positions with employees whose major experience lies outside the environmental professions. Although effort must be made to ensure that these employees understand the sensitivity associated with compliance issues, organizations can use them to successfully manage compliance matters provided they are properly assigned and adequately trained. Furthermore, by emphasizing professional development, organizations can improve the effectiveness of their environmental employees.

4.2.2.2 Fostering Professionalism

Genuine competence in environmental management is achieved by acquiring the skills necessary to understand and respond to the current and future demands imposed by compliance issues. Not only must environmental professionals know the regulatory requirements that apply to their areas of expertise, but they must also have technical skills, such as those required for permitting, reporting, and operating environmental control systems. Furthermore, environmental managers must show good judgment, work efficiently, and communicate well.

Environmental management encompasses several professions, and organizations should strive to emphasize the professionalism of their staffs for several reasons. First, an appreciation for professionalism helps organizations recruit higher quality personnel, because the best professionals prefer working where their talents are respected and applied appropriately. Second, professional behavior by environmental employees creates an atmosphere that encourages respectful communication and diligence in fulfilling their responsibilities. Third, organizations enhance their ability to comply with environmental laws when they entrust tasks that require professional skills to professionals.

Considering that professionalism is highly desirable, organizations should do more to foster the professional development of their environmental talent. The foundation for establishing a professional environmental staff is to hire employees who have the basic skills and credentials generally associated with their positions of responsibility. This is important because it affects the basic competence of the staff, and staffs that are weak on credentials tend to be reluctant to add more qualified persons for fear of undercutting their future prospects. With respect to education, it should be noted that the environmental professions, although varied, are well established and that university-level degrees exist for most professions. Examples of degrees commonly seen in the environmental professions include environmental engineering, hydrogeology, civil engineering, mechanical engineering, chemistry, and biology. This does not mean that these degrees are fungible when it comes to environmental competence. For example, a marine biologist is no more competent to design a wastewater treatment system than an environmental engineer is competent to assess the taxonomic structure of benthic infauna communities. Some people are able to achieve competence in areas outside their formal education but this requires diligent study and substantial experience. Nonetheless, organizations need to carefully assess their needs and hire talent qualified to meet the demands of achieving compliance.

While education forms the cornerstone to becoming an environmental professional, experience and professional credentials are critical. Most professionals need work experience to develop their skills fully. This fact is reflected in the process that states have instituted for registering professional engineers and geologists. To become registered, candidates must first pass the engineer-in-training examination and then work for several years before becoming eligible to take the professional engineers examination. The registration process also incorporates an intense evaluation of the education, experience, and character of the applicants. Because registration reflects a serious commitment to a profession, it is good for organizations to encourage their environmental professionals to obtain the major professional credential associated with their field. For example, engineers should become registered as professional engineers; hydrogeologists, as professional geologists; and environmental scientists, as Qualified Environmental Professionals (QEP). Serious professionals seldom have legitimate reasons for not obtaining the appropriate credentials, especially if they aspire to be promoted to senior management positions.

Another aspect of professionalism is staying informed about current developments. The environmental disciplines produce a great deal of literature in the form of trade publications and academic journals. Considering the rapid change with regard to environmental issues, failing to read this literature can result in professional obsolescence within a few years. It is highly recommended that environmental professionals read at least one environmental science publication with a general orientation, in addition to the major ones oriented toward their specialty. Environmental professionals should also consider reading beyond their particular specialty. The environmental field encompasses several disciplines, and reading literature outside of a person's specialty enables them to maintain a higher level of overall environmental proficiency. Reading broadly will also greatly enhance a professional's insight and ability to respond to subtleties that less developed persons might allow to culminate in noncompliance or inefficient expenditures.

Membership in professional organizations can be a vital part of maintaining an individual's competence. For a modest annual investment, professional organizations provide their members with important technical information and facilitate meeting acquaintances with related professional interests. Because professional organizations provide valuable benefits to their members who, in turn, provide better service to their employers, companies should encourage their professional employees to belong to professional organizations and participate in their activities.

Finally, organizations can go a long ways toward fostering professional development by treating their professional employees appropriately. This means not only treating them with the respect that they deserve but also setting high expectations for their quality of work. Having to check such a person's calculations and ensuring the use of proper grammar may seem trivial and many organizations tolerate substandard performance of this sort, but this sends the wrong message regarding what constitutes professional behavior. By insisting on high quality in all aspects of the environmental function, organizations will markedly improve the performance of their environmental employees and their self-esteem as well.

4.2.2.3 Recognizing and Dealing with Problem Environmental Employees

Most employees with environmental responsibilities are capable and responsible and do not present employment problems; however, those employees who cause problems in the environmental area present special issues not generally encountered in other fields — namely, their actions can expose the employer to exorbitant sanctions. Problem employees can also be difficult to deal with because they are sometimes protected from adverse employment action by whistleblower protection laws. To ensure sustained compliance, organizations must understand the special risks associated with employees that tend to violate environmental laws and act appropriately when confronted with problem employees.

The best way to avoid the difficulties associated with problem employees is to be careful not to place people into positions where they are likely to violate environmental laws. One factor to consider when assigning employees to sensitive positions is their attitude towards compliance. The worst employees to whom environmental duties can be delegated are those who do not respect legal requirements. In extreme cases, resentment towards environmental requirements can motivate an employee to violate requirements whenever he or she believes the misconduct will escape detection. Persons with this attitude expose their organizations to enormous liability, particularly when assigned to duties that involve sensitive regulatory issues such as monitoring or hazardous waste management. Less malevolent but still serious are employees that believe that noncompliance is acceptable so long as it does not harm the environment. The "no harm, no foul" attitude toward compliance can be particularly problematic when applied to sensitive issues such as the integrity of monitoring data or the reporting of releases.

Organizations should also avoid assigning environmental duties to persons who mean well but are not sufficiently diligent to ensure continuous compliance. For example, an employee might diligently inspect and keep records on electrical equipment containing PCBs but violate regulatory requirements by failing to include all the required attachments in the annual documents log. Employees who constantly come up a "nickel short" with regard to compliance activities present a dilemma in that they frequently fail to meet regulatory requirements but often have positive attitudes towards environmental issues. However, environmental protection is not the same thing as environmental compliance, which is why such employees may avoid conduct that harms the environment yet at the same time be willing to violate legal requirements that do not directly affect environmental quality. Because there will always be an element of doubt regarding the completeness of such employees' work, they generally should not be placed in positions where their lack of diligence will probably result in violations.

Another type of problem attitude is seen in employees whose inflated sense of self-righteousness makes them believe that they care more about the environment than their coworkers. These employees tend to view life in terms of the good guys vs. the bad guys and generally perceive businesses that are subject to environmental regulations as the bad guys. Unfortunately, people with such simplistic mindsets often lack the judgment needed to perform environmental responsibilities competently. Moreover, they are frequently passive, indecisive, or indifferent when presented with situations that call for personal action to meet environmental requirements. A notable irony about these employees is that while their tendency toward inaction makes them more likely to cause environmental violations than other employees, their negative attitudes towards their employers make them more likely to report alleged violations to regulatory agencies and other outside groups.

If an organization is to maintain compliance with environmental laws, it must recognize problem employees before their shortcomings lead to significant environmental problems. Although such detection can be difficult, most such employees exhibit certain characteristics that suggest a propensity for causing violations. By watching for these characteristics and devoting special oversight when warranted, companies can avoid assigning compliance-related duties to persons who are unlikely to execute them adequately. Also, the performance of some employees degrades over time, and organizations should appraise performance at timely intervals.

A common characteristic among problem employees is the frequent failure to complete assigned tasks to the standard expected. These employees may show a great deal of initiative when first assigned a task and often show great zeal in undertaking unassigned and unsolicited projects. In fact, it is not unusual for employees that exhibit this characteristic to imply that management is disregarding compliance or safety issues by failing to give priority to their pet projects. Regardless of their apparent zeal and initiative, reliance on this kind of employee to perform even simple compliance tasks almost inevitably results in some degree of noncompliance. A variant is the environmental employee who has a history of sloppy work or poor judgment. These employees manifest themselves by sometimes putting minimal and inadequate effort into their assigned tasks and at other times attempting to carry out their duties in good faith but falling short in important aspects. For instance, such an employee may substantially complete a permit application but omit a few critical items. Although their overall attitude towards compliance may be acceptable, their unwillingness to conform to a high standard can result in widespread shortfalls in meeting compliance requirements.

Another type of problem employee is the kind that refuses to be responsible for their actions. For instance, when queried about the failure to complete a report, such employees may assert that they are not required to do the work or that they were prevented from doing so by management directives or a lack of resources. These employees also tend to blame their failures on coworkers and occasionally may falsely claim that regulatory staff or other advisors have given them assurances that their actions comply with legal requirements. In the worst cases, the workers routinely disparage their coworkers and supervisors. These disparagements tend to encompass general statements alleging incompetence or culpable acts of noncompliance; however, these employees almost never provide concrete examples when asked to support their allegations.

Organizations should be particularly careful when an employee exhibits a punitive attitude toward the company or other workers. Employees with this attitude generally place as their highest priority seeing that punishment or harm befall on those they disfavor. Many such employees will write negative comments in personal journals, memoranda, or operating logs, typically recording distorted descriptions of perceived wrongful acts committed by the employer or coworkers. These employees are especially risky, as they may vent their ill feelings by making false or exaggerated allegations to regulatory agencies or hostile environmental groups. In addition, many of these employees display obvious discontent with either their employer or their position by actions such as withdrawing from compensation plans, openly discussing searches for other employment, and voicing veiled or open threats to sue the employer upon termination or departure.

Employee issues are best handled within the confines of an adequate compliance program. As is explained more fully later in this chapter, an adequate compliance program will incorporate provisions such as corporate policies on compliance, noncompliance reporting mechanisms, and auditing systems intended to ensure compliance. The policy provisions should emphasize the importance of achieving total environmental compliance and spell out that employees are expected to act in ways that ensure compliance. By setting a simple explicit standard for employee conduct, organizations can reduce the difficulties associated with dealing with problem environmental employees. Furthermore, organizations can reduce the potential for noncompliance by designing their reporting and auditing systems to detect problem employees. For instance, the procedures used to review employee performance can be modified to evaluate whether employees are suitable for positions that involve compliance issues.

When problems are detected, they should be dealt with promptly, as even minor misconduct can escalate into serious and widespread violations if allowed to continue. Unfortunately, concern about potential for employee lawsuits has caused many employers to act cautiously when confronting apprehensions about problem employees. This concern is legitimate, but organizations must be mindful that environmental enforcement actions can pose more serious risks than the typical employment lawsuit. Another reason to deal with employee concerns promptly is that federal statutes may protect the employee from retaliation if they have communicated with agencies or management about compliance problems. Because even frivolous invocations of whistleblower protection can greatly complicate efforts to discipline problem employees, it is advantageous to deal with problem employees before they can obtain status as whistleblowers.

If a problem employee does complain to management or a government agency about environmental compliance, the organization must be careful not to act in a manner that could be construed as retaliation for having made such communications. Although expressing concern about compliance issues does not shield problem employees from discipline for misconduct related to other activities, the organization will have the burden of proving that the discipline was not related to whistleblower activities.[40] In cases where it is known that the employee has complained about compliance issues, organizations should not discipline without first consulting with legal counsel, as such discipline has the potential of exposing the employer to a whistleblower lawsuit and in certain cases to charges of obstruction of justice.

Handling problem employees with known or suspected shortcomings with respect to environmental matters is qualitatively similar to the handling of other kinds of employment issues but should be faster and more deliberate. If an employee is generally conscientious but seems unable to carry out assignments with the detail required to ensure environmental compliance, reassignment to nonenvironmental duties may be appropriate. This is particularly true for employees who tend to evade responsibility because they have difficulty understanding environmental requirements but would otherwise be good workers. If an improper act appears to be nothing more than a sporadic or inadvertent omission, then a reprimand or perhaps probation may be appropriate concurrent with enhanced monitoring of the employee's work. Sometimes conscientious employees will fail to meet requirements because of ignorance. Additional training for such employees may be the appropriate response in such instances.

Termination, however, will generally be the best option for employees with punitive feelings toward the organization and coworkers. These employees not only tend to cause violations, but they also frequently convert attempts to resolve problems into contentious and confrontational exchanges. In addition, their punitive mindsets can subtly cause the organization to redirect much of its energy toward fingerpointing and backstabbing. The resulting misery heaped upon the good employees can lead to a long-term erosion of environmental talent as the better employees seek work elsewhere. Organizations that lose this talent will find that the environmental problems are exacerbated and the talent is very difficult to replace. Although termination of an employee is not to be taken lightly, neither should the increased prospect of noncompliance be lightly dismissed.

4.3 SETTING UP AN ENVIRONMENTAL COMPLIANCE PROGRAM

Although corporate compliance programs have their origin in the fallout of massive antitrust and foreign corrupt practices prosecutions in the 1960s and 1970s, they have been increasingly used by organizations to minimize the prospect of misconduct in almost every area of regulation. The overall purpose of such programs is to develop, communicate, and enforce conformance with an organizational ethic against misconduct. By establishing an environmental compliance program, organizations formalize their basis for sustaining compliance and communicate to employees, government agencies, and the public that they are committed to total environmental compliance.

There are other benefits to establishing environmental compliance systems besides their ability to enhance compliance with environmental laws. Under the federal sentencing guidelines, the establishment of an effective compliance programs is one of the few factors that judges may consider for the purpose of mitigating criminal sentences for organizations.[28] However, the government has come to assess the effectiveness of a compliance program for purposes other than criminal sentencing. For example, decisions regarding the appropriate means of enforcement have depended on whether the violating organization had a formal compliance program.[41] By establishing compliance programs, organizations hope not only to avoid noncompliance, but also that government agencies will treat them more leniently should violations occur.

The federal government has defined what it considers to be the elements of a diligent compliance program in the *United States Sentencing Guidelines*.[28] Although these guidelines were drafted in the context of criminal sentencing, they have been widely applied in regulatory settings whenever agencies have need to assess the effectiveness of compliance programs; however, these guidelines also set a benchmark against which the government judges the quality of a compliance program. With regard to what is considered to constitute a minimally effective compliance program, the standards of the *United States Sentencing Guidelines* set forth seven elements:

1. The organization must have established compliance standards and procedures to be followed by its employees and other agents that are reasonably capable of reducing the prospect of criminal conduct.
2. Specific individual(s) within high-level personnel of the organization must have been assigned overall responsibility to oversee compliance with such standards and procedures.
3. The organization must have used due care not to delegate substantial discretionary authority to individuals whom the organization knew, or should have known through the exercise of due diligence, had a propensity to engage in illegal activities.
4. The organization must have taken steps to communicate effectively its standards and procedures to all employees and other agents (e.g., by requiring participation in training programs or by disseminating publications that explain in a practical manner what is required).
5. The organization must have taken all reasonable steps to achieve compliance with its standards (e.g., by utilizing monitoring and auditing systems reasonably designed to detect criminal conduct by its employees and other agents and by having in place and publicizing a reporting system whereby employees and other agents could report criminal conduct by others within the organization without fear of retribution).

6. The standards must have been consistently enforced through appropriate disciplinary mechanisms, including, as appropriate, discipline of individuals responsible for the failure to detect an offense. Adequate discipline of individuals responsible for an offense is a necessary component of enforcement; however, the form of discipline that will be appropriate will be case specific.
7. After an offense has been detected, the organization must have taken all reasonable steps to respond appropriately to the offense and to prevent similar offenses — including any necessary modifications to its program to prevent and detect violations of law.

The considerations in satisfying the first four elements have already been discussed in this chapter. To develop a complete compliance program, organizations need only add the elements regarding monitoring, disciplinary enforcement, and corrective action.

4.3.1 The Monitoring and Detection Function

The monitoring and detection element of an environmental compliance program is intended to encourage compliance by providing oversight of activities that can potentially violate environmental laws. Another important function is to detect situations that can lead to noncompliance so that they can be promptly corrected. Early warning of compliance issues allows them to be corrected at lower cost and effort than happens when they persist and result in serious and sustained noncompliance. To be effective, environmental compliance programs should incorporate, at a minimum, an auditing program and some sort of communication system by which employees may report concerns about compliance in confidence and without fear of retribution.

Two caveats are in order regarding the use of monitoring and detection programs with respect to ensuring environmental compliance. The first is that the monitoring and detection function should not be relied on as a primary means of ensuring environmental compliance. Achieving compliance requires that organizations act affirmatively to establish compliance and not take an approach that depends primarily on detecting and correcting problems. Monitoring mechanisms such as auditing and reporting programs may lead to the discovery of noncompliance, but standing alone will be ineffective at preventing it. The second caveat is that monitoring and detection efforts should not be allowed to escalate to the point where a "police state" environment comes to exist within the organization. Monitoring and detection should be used positively to ensure that activities are performing properly and not as coercive tools.

The effort that organizations need to devote to monitoring compliance varies with the size and nature of the activities subject to environmental regulation. Small operations may be adequately served by occasional auditing and a simple nonretaliation policy to encourage employees to report concerns about compliance issues. Larger operations may be better served by more extensive monitoring. For instance, they may wish to establish an auditing department and implement ombudsmen or confidential reporting services to reduce the possibility that employees may abstain from reporting concerns for fear of retaliation or perceptions that supervisors are not interested.

4.3.1.1 Compliance Auditing

Although an extensive discussion of environmental auditing must be left to other books, an overview of how auditing fits within a compliance program is in order. There are three kinds of auditing commonly encountered in the environmental area. Compliance auditing looks at the conformance of the organization's activities with legal requirements and is the type of auditing contemplated by the *United States Sentencing Guidelines* for the purpose of monitoring compliance. Systems auditing assesses the degree to which an organization is conforming to its environmental policies and procedures but does not necessarily provide information about compliance status. Real property audits focus on assessing whether properties present liability issues by virtue of being contaminated. Compliance and systems auditing have substantial application to ensuring environmental compliance, while real property audits are useful in certain contexts, such as property transactions, but have little utility when used as elements of a compliance program.

Overall, compliance auditing tends to be more effective at ensuring compliance than systems auditing. One reason for this is that compliance audits provide a direct assessment of the endpoint (compliance) and thus do a better job of evaluating how well the organization is actually achieving compliance. Systems audits focus on conformance to procedures and may fail to detect compliance issues caused by employee incompetence or negligence; however, systems auditing can be superior to compliance auditing when it comes to matters such as assessing how well the organization can adapt to changes such as new regulatory programs. In a perfect world, organizations would do both kinds of auditing, and many organizations incorporate elements of both into their audit programs. Nonetheless, organizations with limited resources that want to meet the government criteria for compliance programs should ensure that their audit programs assess compliance.

Compliance auditing generally consists of reviewing operations and assessing whether they comply with environmental laws. Such audits can be done as comprehensive assessments of entire facilities or as focused reviews limited to particular issues. A typical compliance audit is performed by persons independent of the audited facility who assess what compliance requirements apply and whether the facility is meeting those requirements. These audits are usually performed by environmental professionals, but sometimes legal counsel prepares the analysis of legal issues. The format of audit reports can be tailored to meet the needs of an organization's management. Some organizations prefer comprehensive reports that describe each element reviewed during the audit and the associated extent of compliance, while others prefer concise reports that focus on issues that merit further action. The former type has the advantage of documenting the complete compliance status of a facility and can be useful for evaluating trends in compliance matters; however, focused reports tend to be shorter and more to the point, which facilitates comprehension of the issues that tend to be important. Focused reports are also faster and less expensive to prepare.

Systems auditing can provide management with insight regarding how well environmental compliance programs and management systems are functioning, but they have a genuine potential to miss even serious compliance issues. Instead of evaluating compliance, systems auditing looks at whether the organization has developed policies and procedures for managing activities that affect the environment and also evaluate whether sufficient resources have been developed. The significant weakness associated with this approach is that organizations can develop extensive procedures and still fail to ensure compliance either because the employees are not executing them properly or because some substantive issues have been overlooked. Absent a hard look at whether the organization has sufficient talent to ensure sustained compliance, systems auditing can be fooled by the presence of extensive documentation into reporting that the compliance program is functioning well. Unfortunately, the assessment of talent is overlooked or given short shrift in many systems audits.

Nonetheless, systems auditing does have an advantage over compliance auditing in that it can reveal weaknesses in how the organization will react to events that may require reporting or entail other environmental obligations. For instance, a systems audit may reveal that a facility has no procedures to ensure that employees inform the person in charge of the facility that a reportable release may have occurred. Under a straight compliance audit approach, this type of issue may not be revealed until the audited facility has actually failed to report a release. This suggests that even when organizations favor the compliance auditing approach, it is beneficial to incorporate some elements of systems auditing.

Organizations should be aware that environmental audits can enhance their vulnerability to enforcement. Although many government agencies follow a policy of not requesting to see environmental audit reports under normal circumstances, such reports are almost always sought during enforcement actions.[42] Providing such reports can reveal compliance issues that agencies have not previously discovered and also indicate management awareness of issues that suggest that noncompliance is tolerated. Such information can lead to greatly enhanced enforcement. Audit reports can also be discovered by private parties during citizen suits and cost recovery actions. Without a means to protect the confidentiality of audit reports, good faith efforts to uncover and correct compliance problems can greatly increase exposure to legal liability.

To protect the confidentiality of these sensitive documents, organizations can avail themselves of the attorney-client privilege and in some cases an environmental audit privilege.[43,44] In general, the attorney-client privilege provides a better basis for ensuring confidentiality than environmental audit privileges, but it also entails the added expense of involving legal counsel in the audit process. For the attorney-client privilege to apply to environment audits, the audit documents must be developed in the context of a client and attorney relationship, be kept confidential between the client and the attorney, and relate to the purpose of enabling the attorney to provide the client with legal advice. Many states have enacted environmental audit privileges that enable audits to be kept confidential to a lesser extent than the attorney-client privilege. The advantage of relying on environmental audit privileges is that the communications need not be made in the context of an attorney-client relationship; however, the federal government refuses to recognize the environmental audit privileges and it cannot be asserted to prevent the disclosure of audit documents in litigation involving federal agencies. Because these agencies are responsible for a substantial proportion of enforcement actions, the environmental audit privileges have limited practical effect in protecting the confidentiality of audit documents.

4.3.1.2 Employee Reporting Programs

Many regulatory agencies believe that while employees and former employees are not the dominant sources of information about noncompliance, they tend to be very good sources when serious violations are involved. Among the reasons for employees being good sources of information about noncompliance are that they tend to be very familiar with operations at a facility and they can provide comprehensive details about activities that may not conform to environmental requirements. Because employees can be a good source of information, it makes sense for organizations who are serious about achieving total environmental compliance to implement programs by which they can maximize their access to this information.

When it comes to environmental problems, it is almost always better for management to learn sooner than later. Establishing an effective reporting program can lead to the discovery of problems that otherwise could remain concealed for years. Another benefit of reporting programs is that they give employees the option of disclosing problems to internal management rather than external entities, such as government agencies, newspapers, or environmental groups. Organizations generally will have better control over outcomes when they can respond to an internal disclosure. When organizations learn about the problems from external sources, their options will be limited mostly to reactive strategies. For instance, the EPA has a policy under which it waives or reduces penalties if companies disclose violations discovered under the auspices of a corporate compliance program; however, such relief is not available if the government discovers the violation before the corporate disclosure. Thus, companies can retain the ability to seek relief from penalties if an employee discloses misconduct to internal management but will not if the employee reports the problem to a government agency or newspaper.

Although employee reporting programs have substantial benefits, organizations should also understand the associated risks. First, these programs can generate sensitive and discoverable materials. Because the attorney-client privilege will not apply to the typical report, organizations are unlikely to succeed in keeping reports from being discovered in legal proceedings. Second, reporting services can be abused by employees and other parties. One means of abuse is the filing of false reports intended to disparage the company or its workers. Reporting programs can also be abused by employees who request services such as legal advice or personal assistance in matters such as pay raises or performance evaluations. Abuse of reporting programs cannot be avoided altogether but can be minimized. For instance, the company should recognize the potential for false reporting and never assume that allegations are truthful without verification. Although these risks can be substantial and should not be casually disregarded, in most instances the benefits will outweigh the drawbacks.

Many organizations do not make much use of employee information as a compliance tool because traditional management practices have not emphasized open communication between employees and management. Even organizations that have not actively discouraged employees from communicating about compliance concerns may find that their employees are reticent about providing information. As a practical matter, many employees are reluctant to report concerns to supervisors for fear of retaliation. This is particularly the case when supervisors have engaged in improper activities. Even when supervisors are not implicated, employees may be reluctant to speak out against coworkers.

The two basic approaches to employee reporting programs are open systems and confidential reporting services. In an open system, employees openly report to a supervisor or other designated person. The advantage to this approach is that the organization is better able to obtain more information when needed to address a concern. Open reporting is so important that it should be made part of the organizational environmental policy. The considerations in establishing an open reporting system have been discussed in the section on developing policies.

Because some employees will always be reluctant to report concerns openly, organizations can benefit by implementing a confidential reporting service through which employees can report violations without fear of retribution. When feasible, organizations should strongly consider establishing a confidential reporting mechanism in addition to their open communication policies. In fact, it is unrealistic to expect that open reporting will be effective all the time. For instance, an hourly worker could be deeply troubled about being told to dump solvents in a ditch behind a factory but reluctant to discuss concerns with the factory managers because of their possible involvement. However, by being able to contact corporate management in confidence, the worker is able to express all his concerns and the company has retained flexibility in how it responds to the problem. Without such recourse, the employee is more likely to communicate his concerns to outside parties.

For a confidential reporting service to be effective, it is absolutely essential that the confidentiality of persons making calls be preserved, and any attempt to determine the identity of callers will undermine the service; therefore, it is important that organizations affirmatively commit to maintaining confidentiality. It is important to assure employees that the company will not attempt to discover the identity of persons who call the service. One means of doing this is to establish and publish a policy that any attempt to discover the identity of a caller will subject the violating person to discipline.

Not only is confidentiality important to individual employees, but it is also in the company's legal interest to preserve confidentiality. For instance, a breach of confidentiality can essentially destroy a company's ability to conform to the government criteria for corporate compliance programs. Once employees lose faith in the confidentiality of the reporting service, the company's system to detect violations within the organization will be significantly impaired. Without the ability to meet this minimum requirement, the company can then lose its eligibility for leniency under government policies and the *United States Sentencing Guidelines*. Maintaining confidentiality can also help protect organizations from whistleblower claims. In general, employees who call confidential reporting services can assert they are protected by whistleblower statutes and can recover damages if they can prove that the company retaliated against them for making a report; however, such claims become difficult to press if an employee cannot prove that the company knew of his or her identity.

Careful consideration should be made when setting up a confidential reporting service. Confidential reporting services usually consist of some means by which employees can report information in confidence over the telephone; sometimes these services are referred to as "hotlines," but it is better to avoid this term as it suggests that only exigent matters should be reported. It will generally be in the company's interest that all concerns be reported regardless of whether the caller believes them to be major or minor. Although organizations are sometimes tempted to set up simple services where reports can be made via voice mail, it is far better for reports to be received by a person, as the employees who call may not be skilled at communicating and their reports will lack essential details. In these cases, it is useful to have a skilled person available who can assist callers in clarifying what is being reported. It is better to contract with an outside party to operate the confidential reporting service than to have such calls received by company employees, as callers are more likely to leave a means of being contacted if they know that the person receiving the call is not a company employee. In fact, contracting to an outside provider may be the only means of making confidential reporting services feasible for smaller organizations.

The importance of being able to contact those callers willing to leave a telephone number with the reporting service should not be discounted. It is not unusual to receive a complaint and subsequently discover that more detail is needed if the organization is to investigate effectively. For instance, an employee might report that chemicals are being illicitly dumped at a facility, but unless the location is adequately described the organization may be unable to find evidence of dumping. In such cases, the operator of the service can contact the caller in confidence and request additional information. It is likewise common to investigate a matter and find that there is a legitimate explanation for what the caller perceived as misconduct. For instance, an employee could report that a plant keeps operating even when its pollution control equipment breaks down. A quick inquiry reveals that the plant's permit allows such operation. In these circumstances, the operator of the reporting service can contact the employee and provide assurances that the company has investigated the matter and determined that it is operating lawfully.

If the reporting service is to be effective, the organization must inform employees of its availability. It is not sufficient to make a one-time notification. Ideally, the telephone number of the service should be prominently posted in employee break areas. Furthermore, the organization should provide employees with a brief description of how it is staffed and operated to provide them with some assurance that their identities will be kept secret from the organization. For instance, an organization could display posters in break areas that provide the telephone number and explain that an outside contractor operates the service. The posters can also explain that the phone will be staffed at certain hours and that at other times the caller can leave a voice-mail recording.

4.3.2 Discipline

Although discipline is an element of compliance programs that the federal government considers to be very important, organizations should be aware that the government often tends to perceive that punitive leadership is the most effective means of promoting compliance.[15,28] No matter what approach an organization may find to be the most effective, the government's expectation that organizations must enforce their standards through disciplinary mechanisms becomes an essential but thorny element of a compliance program. In general, most employees favor environmental compliance, and many organizations find they can best promote compliance by providing positive assistance to employees. Similarly, the prospect of recriminations and punitive action often deters communication that is needed to address compliance issues. The unthinking imposition of punishment, therefore, can diminish the effectiveness of efforts directed at addressing the most difficult compliance matters.

Fortunately, federal guidelines acknowledge that what constitutes appropriate discipline will depend on the circumstances.[28] Because it is difficult to predict what discipline will be appropriate in most instances that have resulted in violations, organizations should avoid prescribing specific sanctions for violations in a formal policy; instead, determinations about discipline should be made on an individual basis. To succeed in convincing the government that such a policy meets the requirement for disciplinary action, however, it is necessary for the organization to set forth a general framework for discipline.

For instance, organizations that expect that managers will avoid noncompliance by exercising foresight should formally advise them that their ability to ensure that operations abide by environmental laws will be a factor in assessing performance and determining compensation. Similarly, a policy statement oriented toward employee responsibility can state that the company will view in the most negative light those employees who violate environmental laws for reasons of purposeful acts, inattention, or avoidance of personal inconvenience. Guidance such as this can form a foundation for distinguishing between negligent and deliberate conduct in assessing responsibility. It also provides a basis for meting out discipline associated with a compliance issue. For example, an organization that establishes general guidelines for discipline can treat differently a manager who failed to become aware of a permit requirement than one who deliberately disregarded the requirement to meet a production goal. This is important because in some cases the organization may believe that actions aimed at improving employees' abilities to meet compliance requirements may be more effective in the long run than those designed to impose pain or loss. In the former case, the manager could be given more training or informed that his prospects for promotion depend on demonstrating improvement in his or her ability to deal with environmental matters. In the latter case, the organization may want to consider more punitive measures depending on the manager's motivation for deliberately violating a requirement.

Although organizations should avoid constraining their environmental compliance programs by adopting unnecessarily punitive tones, they must nonetheless be quick to discipline strongly those persons responsible for deliberate and serious violations. For example, activities such as tampering with monitoring devices are typically treated by the government as criminal offenses, thus organizations are expected to deal harshly with the employees responsible for those types of actions. For example, termination should be seriously considered for employees who deliberately commit serious offenses, and demotion or reassignment may be appropriate when incompetence or inattentiveness causes serious compliance problems. However, serious criminal misconduct is not the norm for most organizations, and compliance is generally better fostered by imposing less severe consequences. For example, if employees are made aware that having a reputation for meeting environmental requirements will enhance their opportunities for advancement, they will generally work harder at promoting compliance. Also, employees tend to pay more attention to environmental requirements when they know that their efforts at achieving compliance are subject to oversight.

4.3.3 Corrective Action

Government agencies consider how well an organization responds to a compliance problem as one of the most important elements of a compliance program. For example, problems that are promptly discovered and corrected are viewed more favorably than those that are allowed to persist because of indifference or inattention.[8] As a general principle, agencies believe that correcting matters that involve noncompliance should be the organization's foremost priority once discovered. Unfortunately, many organizations are unaware of how strongly the government views the need for prompt corrective action and needlessly escalate the intensity of enforcement they receive.

One reason that prompt corrective action is an element of compliance programs is that the failure to address such problems is consistent with an attitude of tolerating noncompliance. The government is particularly likely to infer that noncompliance is tolerated by management when oversight activities such as audits reveal compliance issues that are not addressed promptly after discovery. Similarly, it is important that organizations respond appropriately to concerns expressed by employees. Although an organization need not presume that employee reports are true or accurate, some sort of investigation should be undertaken to ascertain the merits of the allegations. Such reports should not be left unresolved, even if the final determination is that insufficient information exists to verify the information in the report.

To ensure that problems are addressed promptly, management should emphasize that corrective action must be a top priority when a compliance problem is discovered. If the problem is of a type that can be corrected immediately, then the organization should correct it immediately. If the problem will require time in which to implement a remedy, then the organization should develop, implement, and monitor conformance to a corrective action plan. By addressing problems as quickly as possible, organizations can avoid allegations that they tolerate noncompliance.

An important aspect of the corrective action requirement is that organizations should go beyond merely correcting violations and address the issues that may have contributed to the problem. Upon discovery, organizations should attempt to ascertain what caused the problem and modify the compliance program to prevent a recurrence. Furthermore, organizations should evaluate the occurrence of any compliance matter introspectively. Typical questions that should be asked include why the violation occurred, what could have been done to prevent it, and whether similar problems are likely to exist at other operations. For example, if an audit discovers that a facility has a deficient stormwater pollution prevention plan, the organization should not only evaluate how to correct that deficiency but also the likelihood that other facilities have plans with similar deficiencies. By taking a comprehensive approach when implementing corrective action, organizations greatly enhance their ability to achieve total environmental compliance.

4.4 CONSIDERATIONS IN ADOPTING ENVIRONMENTAL MANAGEMENT SYSTEMS

Environmental management systems differ from compliance programs in that they focus on a establishing mechanisms for achieving a particular environmental goal. While such goals may include environmental compliance, organizations are free to tailor environmental management systems to establish other priorities such as pollution prevention or conserving natural resources. The basic intent of these systems is to enable organizations to manage environmental issues more effectively and efficiently. Although the motivation to implement compliance programs in the U.S. has been driven by the criteria for corporate compliance programs described in the *United States Sentencing Guidelines*, environmental management systems have been developed to encompass concerns beyond compliance.

Common reasons for developing formal environmental management systems are to reduce environmental liabilities, improve the image of organizations, and facilitate cooperative agreements with regulatory agencies. Although implementing a thoughtful framework for environmental management can boost compliance efforts, organizations must understand that such systems cannot succeed in promoting compliance if organizations have not addressed fundamental issues such as qualified staff and the proper allocation of resources. If such fundamentals are not in place, environmental management systems may merely direct organizational activities in a way that allows organizations to violate environmental laws in a more systematic fashion.

Perhaps the best-known framework for implementing environmental management systems is ISO 14000.[45] The framework consists of a set of specifications developed by the International Organization for Standardization (ISO) that have been drafted with sufficient flexibility to accommodate almost any environmental objective. These specifications are based on the Deming model of management which emphasizes the development of clear objectives and the measurement of progress aimed at meeting them. The basic elements of the primary specification, ISO 14001, are as follows:

- Establishment of an environmental policy that expresses the commitment of top management that the organization will comply with environmental laws, control its pollution, and continually improve its environmental management system
- Development of processes to identify the environmental aspects of the organization's activities, ascertain the applicable legal requirements, and define the overall environmental objectives of the organization
- Definition of the roles and responsibilities within the organization for ensuring that the management system requirements are implemented and maintained (this element specifically includes training programs, communication procedures, and documentation requirements)
- Preparation of procedures to monitor environmental performance and to take corrective action when needed
- Establishing a program for periodic reviews of the environmental management system by top management

There is substantial overlap between ISO 14001 and the criteria for corporate compliance programs as set forth in the *United States Sentencing Guidelines*;[28] however, ISO 14001 does not satisfy all the minimum requirements for corporate compliance programs, although it is not inconsistent with them. Specifically, ISO 14001 does not require that the environmental management system be enforced through disciplinary mechanisms nor does it require due care not to delegate discretionary authority to persons the organization knows or should know have a propensity to violate the law. Therefore, management should understand that prudent organizations will need to go beyond ISO 14001 to obtain the full legal benefits of implementing a compliance program.

Environmental management systems can enhance the ability of many businesses to improve their environmental performance, but the decision to adopt a system should not be made without carefully evaluating what benefits the organization expects to obtain by implementing a system such as ISO 14000. One aspect that management should understand is that a specification such as ISO 14000 sets forth a general framework and the organization will be required to put forth great effort in developing a functional system. Furthermore, although the specification may provide a framework that is adequate to improve environmental compliance, it will not necessarily ensure it.

In general, environmental management systems work well only when organizations have addressed the foundational aspects of achieving total environmental compliance. Establishing good leadership is important, as the effectiveness of an environmental management system will depend on a genuine management commitment to environmental performance and a corporate culture that has matured to the point where it understands and responds to the ramifications of environmental requirements. Likewise, the organization must ensure that it has established a core of competent environmental professionals to implement the compliance function. Environmental management systems are effective only when the organization develops procedures that are actually effective at ensuring compliance and is able to execute them consistently. Unless organizations establish a strong foundation for environmental compliance, the adoption of environmental management systems will not cause them to perform significantly better until they address their deficiencies.

5 Ascertaining Environmental Requirements

Identifying the legal requirements that apply to particular operations is critical if organizations are to sustain compliance. In an ideal world, environmental laws would be clear and easily applied to all the activities they are intended to regulate. In reality, environmental laws are frequently ambiguous, complex, and poorly organized, which can make ascertaining their meaning a daunting task. For example, federal and state governments frequently regulate the same conduct and draft requirements that often overlap and are sometimes inconsistent. Even ostensibly clear environmental requirements can defy ready interpretation when applied to individual situations. Moreover, government agencies are sometimes deliberately ambiguous when interpreting their regulations for fear of curtailing their flexibility.[12] Although all this uncertainty makes achieving total environmental compliance difficult, environmental professionals can overcome this problem by understanding the purpose behind environmental laws, how they are viewed by agencies and courts, and how to deal with uncertainty when applying laws to regulated activities.

5.1 BASIC CONSIDERATIONS

Ideally, all employees whose duties include complying with environmental laws would be fully familiar with the requirements that apply to their areas of responsibility. In reality, many persons responsible for environmental compliance must rely almost completely on secondary sources for information about regulatory requirements. Organizations acquire their knowledge of environmental requirements through a variety of means. Common sources of information include trade organizations, conferences, journals, and colleagues. Government agencies also provide information about regulatory requirements through published materials and informal consultations. Although these can all be good sources of information, they also entail the risk that special requirements may be overlooked or that the source may provide poor information. Unfortunately, providing copies of statutes and regulations to persons responsible for compliance does not ensure that the requirements will be understood, as it can be extremely difficult to discern the meaning of many environmental laws solely through their plain text.

If neither primary nor secondary sources of information about environmental requirements are fully reliable, how can parties ensure compliance? One way is to use an sequential approach to learn about environmental requirements. First, environmental professionals should become generally familiar with the overall regulatory scheme before trying to determine the requirements that apply to specific activities. Second, they should learn to be comprehensive when evaluating environmental requirements. For example, because federal and state agencies often regulate the same activities, it is necessary to ascertain the requirements at both the federal and state levels. Similarly, regulatory requirements often reference materials such as other regulations and private standards that must be consulted if total environmental compliance is to be achieved. Third, professionals should be sensitive to pitfalls that impair correct interpretations of environmental requirements such as the practice of some agencies to assign unusual definitions to common terms when drafting regulations.

Understanding the concept of dual regulation is important when evaluating environmental requirements. For instance, some activities may be subject to multiple aspects of regulation, such as air emissions, wastewater discharges, and solid wastes. It is also important to ascertain state and federal requirements separately, as they often impose different obligations even when the same conduct is being regulated. Local requirements should also be assessed, particularly when operations involve activities that are frequently regulated at the local level, such as allowable noise levels and the discharge of wastewater to municipal sewer systems. Another aspect of dual regulation is the profound effect on regulated activities that agencies can exert by their oversight and approval responsibilities. For instance, the U.S. Environmental Protection Agency (EPA) maintains oversight authority when it delegates authority to a state to administer federal environmental laws such as the Clean Air Act. This means that even when such matters as permitting are the primary province of a state agency, the federal government retains considerable authority and political leverage to influence state decisions.[44] Similarly, states sometimes retain substantial power to influence federal actions. For example, states retain authority under the Clean Water Act to certify wastewater and dredging permits issued by federal agencies.[46] Therefore, when dealing with environmental matters that may be regulated by federal and state agencies, it is important to address the concerns that may be raised by the agencies with the oversight or approval responsibility.

Environmental laws, like other regulatory requirements, can be difficult to interpret, and it is strongly recommended that attorneys be consulted when an erroneous interpretation might result in a serious compliance situation. One reason for this is that attorneys are trained to construe the meaning of laws and how they are applied. This does not mean that other sources of information about legal requirements will necessarily provide incorrect assessments, but the probability may be higher. Another reason to consult with counsel on sensitive issues is confidentiality. Communications made to attorneys during the course of legal services are protected by the attorney-client privilege, which means that they are protected from disclosure to government agencies and private litigants. No such confidentiality attaches to communications made to other common sources of information about environmental requirements.

Even though the resolution of difficult issues involving determinations of regulatory requirements should be made with assistance of counsel, environmental professionals need to understand the basic considerations that affect the interpretation of environmental laws. Environmental professionals will find that being familiar with the general principles of statutory and regulatory interpretation is useful in understanding how uncertainty affects the application of legal requirements. Likewise, understanding how judicial attitudes toward environmental laws and agency actions affect the implementation of regulatory programs will help professionals put specific requirements into a proper perspective. Even though appreciation of the law and its meaning is essential, environmental professionals also need to understand how the law is actually applied. For instance, one impediment to achieving compliance is the failure to understand that government agencies do not have the power to waive compliance to written requirements. In addition to understanding the principles by which laws are interpreted, it is also important to understand the concept of enforcement discretion.

5.2 THE DIFFERENCE BETWEEN LEGAL REQUIREMENTS AND ENFORCEMENT DISCRETION

The concept of enforcement discretion has important implications to environmental requirements and is sometimes a source of confusion to those responsible for ensuring environmental compliance. Although everyone is supposed to obey the law, government agencies are not legally required to enforce against every violation they discover. The power to choose whether to pursue or forego enforcement is known as *enforcement discretion*. The ability to exercise enforcement discretion is important to agencies, as it allows them to allocate their resources as they see fit and to act in a manner that appears just and proper under the circumstances. Enforcement actions can consume enormous resources, and government agencies must decide how to direct their enforcement activities in the face of budgetary constraints. This means that some violations will not be enforced because the agency chooses to devote its resources to other issues. In addition, issues of simple fairness may convince an agency that enforcement is not warranted. For example, an agency might consider enforcement unnecessary when it discovers minor violations that have been promptly corrected.

Although the availability of enforcement discretion is something that benefits regulated parties, they should not expect that agencies will be merciful when they discover violations, because the enforcement function is a major part of a regulatory agency's mission and should not be taken likely. Agencies are established to promulgate and enforce environmental laws, and organizations that rely to any degree on the expectation that agencies will not punish violations will inevitably incur enforcement. Furthermore, the power to exercise enforcement discretion is vested within the agency, and there are few legal limits on its exercise. In other words, courts almost never interfere with an agency's decision to pursue or decline enforcement. The fact that a regulated party believes that a particular enforcement action is grossly unfair does not entitle it to seek redress from a court. Similarly, a party that feels it has been unfairly targeted for prosecution cannot sue the government to force it to commence similar actions against its competitors. Simply put, there is no private right to either constrain or compel the government's exercise of enforcement discretion.

Unfortunately, many organizations have come to assume that they can rely on the government to act fairly or to continue exercising discretion according to its prior practice. This assumption, as many have discovered, may not hold up over time. Fairness in enforcement matters is in the eye of the beholder, and the fact that an organization may perceive that a violation is reasonable will not protect it from enforcement by an agency or private citizens. Likewise, changes to an agency's resources, priorities, or political position may alter its approach to enforcement.[47] Changes in position regarding enforcement can also occur at the personnel and inter-agency levels. For example, a newly assigned employee at an agency may decide to pursue enforcement of violations that the former holder of the position chose to ignore. Similarly, the EPA may choose to exercise its parallel enforcement powers to pursue matters that a state agency has failed to enforce.[44] Therefore, the fact that an agency is aware of a violation and has chosen not to proceed with enforcement provides little assurance to a regulated party that it will avoid enforcement in the future.

Organizations should consider that enforcement discretion can allow them to resolve ambiguous or otherwise difficult compliance issues without incurring sanctions. Furthermore, they may want to consider how best to manage their compliance activities to increase the chances that agencies will forego enforcement should contested issues arise. One way to encourage favorable treatment in situations where compliance is questioned by agencies is to establish a reputation for vigilance and integrity when it comes to environmental compliance. Organizations that aggressively skirt the edges of environmental law become poor candidates for enforcement discretion, while organizations that favor deliberate and conservative approaches to achieving compliance issues will find that agencies are far less motivated to pursue enforcement. Agencies also tend to be impressed with organizations that implement environmental compliance programs and hire senior environmental officers with strong reputations for competence and integrity. A sophisticated environmental staff also makes a good impression. Although the possibility of enforcement discretion should not be a cornerstone of an organization's compliance program, it should not be ignored either. By implementing a program intended to achieve total environmental compliance, organizations can increase the likelihood that agencies will give them the benefit of the doubt in a complex and highly regulated area.

5.3 THE PUBLIC WELFARE DOCTRINE AND HOW COURTS VIEW ENVIRONMENTAL LAWS

Environmental laws fall under the general category of public welfare laws. These laws are intended to protect the public at large rather than individuals or members of a limited class. Because the purpose behind public welfare laws is to promote the public good, they are applied differently than laws intended to protect private or commercial rights. First, courts tend to interpret public welfare laws broadly when necessary to effectuate their intent. The rationale behind construing these laws liberally is that the promotion of the social good should not be obstructed through an overly strict reading of the law. The public benefit is that courts will not allow irrational loopholes or overly strict interpretations asserted by regulated parties to prevent the government from remedying problems that the laws were enacted to address. The drawback to regulated parties is that it becomes more difficult to determine the precise limits that laws impose on conduct when they are construed in this manner.

Although liberal construction can sometimes make laws appear more rational, organizations should never assume that liberal construction will be applied to authorize conduct apparently prohibited by a law that seems irrational or unwarranted under the circumstances. As a general rule, courts and agencies are far more likely to rely on liberal construction to extend the reach of government power and will rarely apply it in ways that allows parties to avoid unnecessary or irrational regulation. In other words, the concept of rationality in environmental regulation is basically a one-way street; it may extend government power when necessary to safeguard public welfare but will not excuse compliance with requirements that are unnecessary to protect the environment.

There is a limited exception to the standard practice of construing environmental laws liberally. Strict construction is sometimes applied when laws do not provide fair warning of the conduct expected of regulated parties before imposing liability on them. Unfortunately, it can be difficult as a practical matter to ascertain which laws will be construed liberally to expand government power and which will be construed strictly to curtail the imposition of sanctions. Although there have been several cases where courts have ruled against the government on the grounds that regulatory requirements were not clearly expressed, there have been many other cases where courts have favored agency interpretations of ambiguous requirements.

The practice of courts deferring to agencies can have a profound effect on the way that environmental laws are applied and enforced. To understand why courts tend to defer to agencies, it is helpful to first understand the role of courts in environmental regulation. Courts exist primarily to resolve disputes over how the law is applied. Courts do have the power to create law in some areas such as contracts or torts but cannot fashion laws that conflict with statutes or regulations. Because most environmental laws are of statutory or regulatory origin, the role of the courts is mostly to interpret the law rather than create it. As a result, the focus of most adjudicated disputes will be on what a law was intended to mean or whether agencies followed procedural rules associated with rulemaking or agency hearings. Issues concerning environmental policy or the necessity of regulatory requirements are rarely considered by courts. This means that courts do not focus on whether an agency's decision was good or bad but whether it acted within its authority. For example, a court would refuse to consider a defense to a hazardous waste storage violation based on the contention that a 90-day limitation on storage without a permit is an unnecessary regulation.

When deciding environmental matters, courts tend to defer to agencies with regard to legal and technical issues. This tendency has profound consequences on regulated parties, as it creates a standard of judicial review that is markedly skewed toward government agencies.[48] The willingness to defer on legal issues can be attributed to a belief that agencies are in the best position to know the meaning of the regulations they promulgate and the laws that legislatures have charged them to administer. In the absence of proof that the agency actions had no rational basis or clearly were outside its scope of authority, courts are likely to find that the agency's construction of a legal requirement is permissible. Under this "arbitrary and capricious" standard, a regulated party cannot prevail in a dispute merely because it has a better or more reasonable position than the government agency. Instead, the regulatory party must show that the agency had no rational basis for acting or deciding as it did. Even if a court determines that the agency did not make the most reasonable decision, it will still uphold the agency's position absent a finding that the agency action was entirely arbitrary.

The willingness to defer to agencies with respect to technical matters is based on the belief that courts lack the expertise to evaluate the technical considerations involved in promulgating regulations. Therefore, a judicial doctrine has developed that courts will defer to an agency's decision in matters involving technical issues unless the regulated party can demonstrate that the decision had no rational basis at the time it was made. Although courts are not required by doctrine to defer to an agency's technical position when facts are contested at an enforcement hearing, they often do so because of the difficulty of judging the merits of competing expert assessments.

Another aspect of environmental laws that limits the ability of regulated parties to defend themselves in enforcement actions is the limitation of judicial review of regulations and permits. Most environmental regulations and permits must be challenged shortly after they are established or issued or else judicial review is precluded. This practical effect of this limitation is that regulated parties cannot contest the validity of legal requirements during enforcement actions. For example, a party that objects to a permit provision that restricts how hazardous wastes may be stored cannot wait to see how the agency interprets or enforces the provision, because by the time an enforcement action arises the period for commencing judicial review will have expired.

5.4 DEALING WITH REGULATORY UNCERTAINTY

Environmental laws are the expression of how the government chooses to regulate conduct that may affect the environment. Because these laws involve complex issues of science, technology, and policy, they can be difficult to formulate and even more difficult to interpret and apply to specific circumstances. The complex nature of environmental laws also increases the risk that regulated parties will fail to comply because they misunderstand their meaning or applicability. In many cases, it will be impossible to know with absolute certainty what a particular environmental requirement means until a court or agency makes a definitive interpretation.

To make a difficult situation worse, it is not always obvious when environmental requirements are unclear. Sometimes agencies express environmental requirements in ways that are ambiguous, incomplete, or almost deceptive. Although these problems are caused by the way government agencies choose to draft regulatory requirements, the failure to grasp the intended meaning of poorly structured regulations can cause regulated parties to unwittingly commit pervasive and serious noncompliance. However, the risks presented by bad laws can be reduced by recognizing how poor usage and tenuous applicability encourage incorrect interpretations and how to reduce the possibility of misunderstanding the legal requirements that apply to an organization's activities.

That interpreting environmental requirements can be a difficult proposition is evidenced by the many court cases involving environmental issues. In fact, it is not unusual for even experienced environmental attorneys to differ on what requirements are imposed by various environmental laws. As noted above, it is not just the immense volume of environmental law that makes interpretation difficult; the complexity associated with drafting text to regulate innumerable activities that affect the environment can make discerning meaning a formidable challenge. Attorneys practicing environmental law draw upon many skills acquired through education and experience to advise clients on how the law applies to them. Understanding how requirements apply to specific situations may involve numerous sources such as statutes, regulations, court decisions, administrative history, legislative documents, constitutions, evidentiary rules, and proposed authority. Although the need for legal skills makes the use of counsel necessary when dealing with complex, ambiguous, or obscure requirements, as a practical matter the regulated community cannot rely on attorneys to inform them of every legal requirement they are obligated to follow. Nonetheless, understanding basic principles of ascertaining environmental requirements can help environmental professionals make better decisions when dealing with compliance issues.

The first principle to follow when ascertaining an environmental requirement appears straightforward but is often neglected — read the text of the requirement carefully. Considering the consequences at stake, it is incredible how frequently organizations participate in regulated activities based on incomplete or strained readings of the associated regulatory requirements. Common examples include completing permit applications improperly, failing to fulfill training and inspection requirements, and availing themselves of exemptions that do not apply to their activities. In many instances, such mistakes can be avoided by careful reading. It is also critical to read all the relevant material. For example, if an industrial pretreatment permit requires the discharger to comply with the local sewer ordinance, the discharger should obtain and read the ordinance. Similarly, if a wastewater discharge permit requires analyses to be conducted in accordance with 40 C.F.R. Part 136, the operator of the facility should become familiar with the additional requirements that are imposed by those regulations.

When attempting to ascertain the meaning of an environmental law, it may become necessary to consult sources other than the text of the requirement in order to clarify what the agency is trying to express; therefore, it is helpful to understand what kinds of secondary sources may be available to help in understanding a requirement. Most environmental requirements are imposed either in the form of regulations or permit conditions; however, before an agency can issue a regulation or permit, they must follow certain procedures and create an administrative record. With regard to rulemaking, agencies will generally create a file or docket in which the important documents supporting the agency's action are kept. The administrative record underlying a regulation may incorporate development documents, fact sheets, and descriptions of proposed rules such as those often seen in the *Federal Register*. The administrative record will also contain comments submitted by the public regarding proposed regulations and the agency's responses. The permitting process also creates an administrative record that in many respects is similar to the records developed for regulations. Because administrative records generally contain more explanatory material than the final regulations or permits they support, they often provide information that is useful in determining how requirements are intended to apply.

Another source of secondary information about the meaning of environmental requirements can be found in interpretive material issued by agencies and in court decisions. With respect to interpretative materials, agencies are bound by their official interpretations of environmental requirements. This means that policy statements, formal guidance, and program memoranda serve as good sources of information regarding what a requirement means. However, less reliance can be placed on unofficial representations such as those made by lower level employees. Although agencies are bound by their official interpretations, they do not necessarily carry the force of law should a regulated party disagree with the substance of the interpretation. For example, interpretations cannot be used to impose requirements that go beyond the scope of the regulation or permit being interpreted.[49] As discussed above, though, courts are highly deferential to agency interpretations, and parties who disregard agency interpretations do so at great risk.

As noted previously, legal counsel should be consulted when organizations are unsure how environmental requirements apply to their operations, but organizations can improve their ability to comprehend environmental laws if they understand the common reasons why such laws are misunderstood. These reasons include poorly drafted textual provisions and the mechanisms by which the laws are implemented.

Awareness of these obstacles to comprehension can also alert organizations to situations where they need to seek more information about what a legal requirement means.

5.4.1 Unusual Meanings

Legislators and regulators seem to delight in attributing unusual meanings to common words and phrases when drafting environmental laws. A notable example is the definition of "solid waste" in the Resource Conservation and Recovery Act (RCRA) which encompasses "solid, liquid, semisolid, or contained gaseous material."[50] Laws that use common words to express contorted definitions impede the ability to achieve compliance because they lack the ambiguity that signals the need to obtain an expert opinion on what the law means. Legally speaking, the government is not supposed to establish regulations that trap the unwary through the use of unusual meanings. In some instances, regulated entities have convinced courts that they should not be penalized for violating requirements that an agency intended to impose but did not adequately express; however, the reality is that many organizations who unintentionally violated confusing or inarticulate laws end up paying significant penalties either because courts tend to defer to an agency's interpretation of a law or because settlement in the face of legal uncertainty appears to be the best course of action under the circumstances.

Avoiding traps created by unusual meanings is difficult, as the text generally does not indicate that the agency may have intended something other than the apparent meaning of the regulatory provision. There are a couple of practices, though, that can be taken to reduce the chance of being caught unaware of the intended meaning. The first is to make a standard practice of reading the definitions sections of statutes, regulations, and permits as the first step in discerning their requirements. Although the definitions are sometimes tedious, they can alert the reader to terms that are used in nonstandard ways. The second practice to develop a sensitivity to intended meanings that are camouflaged in simple but distorted English is to read broadly on the subject of environmental regulation. Reading journals and texts is not a particularly efficient way of learning about poorly expressed requirements, but it should already be a practice of every environmental professional, and paying attention to regulatory issues does not add much to the burden of professional development.

Case Study: 40 C.F.R. § 122.41(m)(2)

Almost every wastewater discharge permit in the U.S. contains a provision that seems on its face to allow operators more flexibility to maintain treatment systems than the government is willing to allow. The maintenance bypass exception appears to allow operators to bypass treatment systems when necessary for essential maintenance:

"***Bypass not exceeding permit limitations.*** *The permittee may allow any bypass to occur that does not cause effluent limitations to be exceeded, but only if it is also for essential maintenance to assure efficient operation.*"

The EPA's intended meaning, though, does not conform with the apparent scope associated with the terms "essential maintenance" or "efficient operation." The use of these terms in the regulation conveys the impression that bypassing treatment works is permissible when maintenance can be done without exceeding effluent limits and is a necessary component of enabling treatment systems to operate at optimal performance. Dictionaries define "essential" as something that is necessary or fundamental and "efficient" as producing the desired effect with a minimum of effort. Moreover, the EPA regulations do not define either term differently than their common meaning. Unfortunately, the EPA construes the maintenance bypass exception in a way that fails to embody the concepts of essentialness or efficiency. Under the EPA's interpretation, the maintenance bypass provision applies only to emergency maintenance:[51]

"*Generally, maintenance is that which is necessary to maintain the performance, removal efficiency, and effluent quality of the pollution control equipment. However, for purposes of this section, it is necessary to distinguish between maintenance that is essential and that which is routine. Further, a distinction must be drawn between what is considered essential maintenance for industrial treatment systems and that for publicly owned wastewater treatment plants (POTWs). ...Maintenance that can be performed during periods of nonprocess operation at an industrial treatment facility is considered to be routine maintenance, not essential maintenance. However, repairs and maintenance that cannot wait until the production process is not in operation would be deemed essential. ...*

"*Unlike most industrial facilities, POTWs are required to operate continuously. Therefore, maintenance must normally be conducted while the treatment facility is in operation. In this situation, it is often unavoidable to bypass certain equipment during maintenance. These maintenance activities would generally be classed as essential. However, since POTWs frequently have capacity exceeding normal loading, maintenance can normally be conducted during periods of lower flow with no loss in treatment plant performance.*"

Under the EPA's strained interpretation, routine maintenance is not essential maintenance, despite the fact that competent operators know that routine preventive maintenance is essential to ensuring efficient long-term operation. In addition, under the EPA's interpretation, the maintenance bypass provision applies differently to industrial and municipal facilities, although no such distinction exists in the text of the provision.

The EPA could have easily clarified the maintenance bypass provision by using words that better express the meaning it intended to convey. For example, the word "emergency" could have been a better choice than "essential" if the EPA intended to exclude routine preventive maintenance from the exemption. Likewise, use of the word "repair" instead of "maintenance" would convey what actions are allowed under the bypass exception. The use of simple and precise words would be a far more conscientious way of setting forth a regulatory requirement than publishing an interpretation in a 1984 issue of the *Federal Register*, where the persons most responsible for compliance are unlikely to see it.

5.4.2 Ambiguity

Another troublesome area of environmental regulation is the ambiguity associated with environmental requirements. Typically, ambiguity occurs because regulatory terms are written in ways such that they cannot be applied with certainty to conduct that may be regulated. A good example of definitional ambiguity is the term "point source" as used in the Clean Water Act. Point sources are defined by the Act as discrete conveyances from which pollutants are discharged to surface waters.[52] Such a definition works well when applied to common scenarios such as effluent pipes from factories or municipal storm drains. Because point sources must have permits issued pursuant to the National Pollutant Discharge Elimination System (NPDES) program, whether a particular discharge of pollutants constitutes a point source is a critical regulatory determination. Unfortunately, determining what constitutes a point source becomes more difficult outside the context of the traditional discharges that Congress contemplated when enacting the Clean Water Act. For example, courts have held that dropping of bombs from aircraft into the Atlantic Ocean constitutes a point source requiring an NPDES permit, but a person throwing a vial of HIV-contaminated blood into the Hudson River is not a point source.[53,54] Courts have yet to decide whether dropping vials of HIV-contaminated blood from aircraft would constitute a point source or whether persons can throw bombs into the Hudson River without running afoul of the permitting requirements of the Clean Water Act. Irrespective of what courts may decide, ambiguous environmental laws present a real problem to those who are responsible for ensuring compliance.

Agencies can also make regulations ambiguous by omitting clarifying text or by creating a requirement by implication. For example some states have promulgated regulations that require operators of air emission sources to record and report upsets which cause air emission limits to be exceeded. Although such requirements might imply that sources may lawfully continue to operation if the upset is beyond the reasonable control of the operator, they do not actually speak to this issue. Absent explicit language or an administrative history setting forth an upset defense, these upsets are probably enforceable. Nonetheless, such ambiguity with respect to important issues, such as a whether a facility may operate during an upset, can lead to difficult decisions and can encourage poor judgment. By omitting text that clarifies whether a facility may or may not lawfully continue operations during an upset, states unnecessarily fail to communicate their expectations clearly.

Sometimes legislatures and agencies create ambiguous laws because it is difficult to craft clear provisions. In many instances, the regulated aspects of conduct are difficult to define, and this difficulty is reflected in mushy text. Sometimes proposed legislative and regulatory provisions are hotly contested, and ambiguous language is used to avoid or resolve disputes during the legislative and administrative process. Another cause of ambiguity is that the promulgating agency is unfamiliar with the practical aspects of the operations to be regulated and therefore omits text that will be needed if the regulations are to be applied sensibly. In addition, some people believe that agencies consciously make regulations ambiguous to encourage some degree of noncompliance so they have leverage to exert over regulated parties or to preserve their flexibility to reinterpret regulations in the future.

Regardless of the cause, regulated parties must resolve ambiguity in environmental requirements if they are to ensure compliance. Often the better way to deal with ambiguity is to prevent it by participating in the rulemaking process. Unfortunately, the preventive approach is not an option when faced with legal requirements that are no longer open to comment or judicial review. In such cases, as discussed later in this chapter, ambiguity can sometimes be resolved by resorting to sources other than the text of the regulatory provision. Even when efforts to get the agency to clarify a requirement fail, organizations can consider changing their operations or procedures in ways that ensure compliance by avoiding the regulatory issue. For example, a permit may specify that monitoring be done using "Method X or its equivalent." Although it may be unclear whether Methods Y or Z are regulatory equivalents, the use of Method X for monitoring will avoid the issue altogether.

5.4.3 Regulatory Gaps

A frustrating problem that occurs with particular frequency with complicated regulations are gaps in regulations which make it difficult to determine how the regulations address specific situations. Some gaps reflect situations where a regulation seemingly should apply but apparently does not. For example, the EPA amended the RCRA regulations in 1984 to incorporate a satellite accumulation provision that allows generators to accumulate up to 55 gallons of hazardous wastes near the point of generation for any length of time.[55] The agency, however, did not discuss how this provision applies to drip pads when these units became regulated in 1990, although many drip pad operators generate less than 55 gallons per quarter of hazardous waste and typically store it in a single container.[56] Because this is the specific kind of situation which the satellite accumulation rule was intended to address, the failure to address satellite accumulation creates a gap where the allowable conduct becomes less certain.

Another kind of gap is created by legal provisions that mandate compliance with procedures that do not exist. These gaps are not always apparent when viewed abstractly but can become painfully evident when a situation occurs that is not well covered by the requirement. Wastewater discharge permits typically have provisions that require the use of approved analytical methods; however, these provisions can only be complied with when approved methods exist for all the parameters for which monitoring is required. Similarly, a permit can require the use of a monitoring procedure whose terms make it inapplicable to the sampled matrix and direct the user to use another method. In such cases, it will be unclear whether the permittee is required to use the specified method even though it is technically unsuited to the matrix or whether they should use the referenced procedure which may not provide the data in the form the agency needs.

The safest way to deal with regulatory gaps is to take conservative approaches regarding compliance. Given a choice between a 90-day time limit or a 55-gallon quantity limit for the accumulation of hazardous wastes removed from a drip pad, the 90-day time limit is the more conservative limit from the perspective of ensuring compliance. Similarly, careful review of proposed regulations and draft permits can reveal issues that need to be resolved before the final product is issued. Long-term approaches to resolving apparent regulatory gaps can consist of obtaining formal interpretations from agencies or requesting changes in regulations through rulemaking petitions.

5.4.4 Stays of Regulatory Requirements

Stays are orders issued by government agencies and courts that negate or modify regulatory requirements. The most common reasons for agencies to issue stays are to avoid burdensome effects of regulations pending a court challenge or to mitigate aspects of regulations that were not contemplated at the time they were promulgated. For example, the EPA partially stayed the listing of wood-preserving wastewaters in 1991 in order to avoid classifying uncontaminated wastewaters such as noncontact cooling water as hazardous wastes pending amendment of its hazardous waste regulations.[57] Although administrative stays are typically published in administrative registers, they generally are not published in other compilations such as codes. Unless parties are aware of stays, they may unwittingly be influenced by regulatory text that is not currently in effect.

Stays are also issued by courts during cases filed to determine the legal validity of statutes and regulations. Courts may issue temporary stays pending the resolution of a contested legal requirement to avoid undue hardship on the affected party and may issue longer term stays when justified by the resolution of the matter. Judicial stays have varying degrees of applicability. For example, some stays may be limited to the parties involved, whereas others may apply to the regulated community more broadly. Generally, the extent of the stay must be determined by looking at the ruling through which the stay is issued. Some stays become publicly available when the court decides to publish its decision, while other stays may go unnoticed by the public if the court or parties do not publicize them. As is the case with administrative stays, judicial stays can easily go unnoticed by parties who do not follow regulatory developments closely.

Considering the difficulty of directly monitoring the activities of agencies and courts for the issuance of stays, most organizations will find the best means of keeping informed about stays is to belong to a trade association that follows these developments. National associations often do a good job of monitoring developments that affect federal laws, but membership in a local or regional association may be necessary if a regulated party needs information about state actions that affect regulatory requirements.

5.4.5 Regulatory Interpretations and Guidance

Agencies frequently issue interpretations that explain how regulatory requirements are to be implemented. These materials can assist regulated parties in interpreting legal requirements but also have the potential to alter the substantive effect of regulations in ways that were not disclosed or even intended during the original rulemaking. Such interpretations and guidance are not laws in the sense accorded to statutes or regulations but are generally upheld by courts unless they stray too far from the regulatory text they purport to interpret. As a legal matter, agencies must use formal rulemaking procedures when they want to impose new requirements onto regulated parties. These procedures entail formal publication and provide the public with the opportunity to provide comments before their promulgation in final form. They also provide a mechanism by which aggrieved parties have the opportunity to challenge rules in court. Interpretations are subject to none of these requirements, and agencies may use them only to clarify or explain preexisting requirements.

It is important to understand that most employees at agencies do not have the authority to issue interpretations that bind the agency.[34] To be binding, interpretations must be issued by a person with sufficient authority to speak on behalf of the agency. In general, such authority will lie only in the higher levels of agency management. For example, a statement signed by the administrator of an agency that sets forth how the agency interprets a specific statute will constitute the type of interpretive guidance that will bind the agency. An off-the-cuff remark from the agency's air-permitting specialist that a proposed electric power station need not consider emissions of nitrogen oxides in its application for an emissions permit would not constitute the kind of guidance that a regulated entity could rely on, as the statement changes a preexisting regulatory requirement and is made by an employee who lacks authority to bind the agency on such matters.

Agencies may issue interpretive material in several forms. The preambles to regulations are important sources of interpretive guidance, as are manuals and directives. Some agencies assemble correspondence into administrative compilations that represent their positions on regulatory issues. Such materials can be invaluable tools for ascertaining regulatory requirements but must also be used with some degree of caution. For example, many agency documents contain information that has been disclaimed by the agency as having no legal effect. This means that should a party rely on an interpretation in such documents as a defense to an enforcement action, the agency will assert that the interpretation is nonbinding. Also, mistakes in interpretation may be corrected without notice to affected parties. For example, agencies that maintain compilations may remove documents that they find to contain erroneous statements or repudiate them in subsequent additions to the compilation. This practice can expose organizations that rely on faulty interpretations to enforcement liability, particularly when documents containing such interpretations are quietly deleted or repudiated

One reason that agencies issue interpretations is to put regulated parties on notice with respect to how the agency intends to apply a legal requirement. Sometimes interpretations are requested by regulated parties to clarify terms used in regulations or to ascertain how a regulation applies to particular situations. By incorporating interpretations and guidance into compliance decisions, regulated parties can bolster their efforts to achieve total compliance. Similarly, the failure to consider such materials can lead to misunderstandings and cause noncompliance.

Another reason agencies issue interpretations is to stretch the meaning of existing regulations to avoid having to issue new requirements. Although agencies are not supposed to constructively amend regulations by reinterpreting them in ways that substantially alter their original meaning, the burden and time required by formal rulemaking sometimes tempt agencies to "interpret" new requirements they assume that regulated parties will follow.

Although courts tend to defer to agency interpretations, they occasionally reject those that conflict with the plain meaning of the regulatory requirement. When the U.S. Occupational Safety and Health Administration (OSHA) first established its Hazard Communication Standard, it explicitly exempted wood and wood products. As evidence became available that wood dust could cause some adverse health effects, the agency issued an interpretation that wood dust was not covered by the exemption. The basis for this exemption was that workers were generally not as familiar with the risks associated with wood dust as they were with solid wood products and also that wood dust was not a commercial product. However, this interpretation ignored the fact that wood dust is made out of wood (an exempt substance) and that wood dust is in fact marketed as fuel, fillers, and raw material for making products such as particleboard. When an administrative law judge ruled against OSHA in an enforcement proceeding, the agency later used formal rulemaking to amend the Hazard Communication Standard.[58] Unfortunately, courts generally will not rule on whether an agency interpretation is correct until the agency actually enforces the interpretation. This means that parties who want to act contrary to an interpretation will have no opportunity to challenge it until the agency commences an enforcement action.

Regardless of the motivation of formal agency interpretations and guidance materials, environmental professionals should become familiar with the major documents that pertain to the regulated fields encompassed by their professional responsibilities. For example, persons whose responsibilities include obtaining permits for air emission sources should be familiar with the EPA's *New Source Review Workshop Manual*, as this document sets forth the agency's view of how the new source review programs are supposed to function.[59] Unfortunately, agencies vary in the extent to which they make regulatory interpretations and guidance available to the public. Some agencies widely publicize the availability of interpretive and guidance materials, while others make little effort to make supplemental administrative materials available to the public.

Organizations are often tempted to call environmental agencies to determine what requirements apply to them, but such consultations can be risky for several reasons. One problem is that the organization may not communicate the full nature of the situation to the agency, and the resulting advice will not reflect all the relevant concerns. Furthermore, although most agency staff are knowledgeable about the requirements within their area of responsibility, they do not have the authority to bind the agency to a legal position. This means that if a regulated entity relies on erroneous advice it will not be protected from enforcement.

Nonetheless, there are times when consulting with agency staff makes sense. For instance, agency staff are often the best contacts for determining whether a proposed technical action will satisfy a requirement expressed in general terms, such as when a permit requires a facility to take reasonable measures to control fugitive emissions. In many instances, consultation and approval by agencies will be legal requirements. Common situations where approval is required or at least advisable include the closure of hazardous waste management units, selection of continuous monitoring equipment for air emissions sources, and modeling to support applications for environmental permits. Another area where agency staff may need to be consulted concerns the interpretation of permit conditions that are specific to a particular facility and not readily interpreted from other regulatory materials. In such cases, agency staff may be the only conclusive means of determining the regulatory intent of these provisions.

Sometimes it becomes desirable to request a formal interpretation of a requirement from an environmental agency. Few agencies have provided procedures for requesting formal interpretations, and the effectiveness of such requests is often a hit-or-miss proposition. Furthermore, there are no guarantees that a request for an interpretation will result in an adequate explanation or one that the regulated party feels is reasonable or even tolerable. In some cases, there is a risk that the inquiry will alert an agency to serious compliance problems. If a party decides to request a formal interpretation of a requirement, it should describe the factual situation in sufficient detail to enable the agency to make an informed interpretation. In addition, the request should be framed in the form of specific questions to discourage vague responses from the agency. Because low-level employees will lack the authority to bind the agency to a particular interpretation, a request should be addressed to a senior agency official with responsibility for the program that regulates the activity for which the interpretation is sought. Organizations that wish to preserve their confidentiality can make requests through an intermediary such as an attorney.

Although formal interpretations can greatly assist in ascertaining and clarifying legal requirements, organizations should realize that a response from an agency may not always adequately clarify an ambiguity. Sometimes the responses are ambiguous and do not provide the degree of clarification needed to provide comfort. Furthermore, it is not unusual for a senior official to direct the request to a subordinate whose response may carry less weight in binding the agency to the interpretation; therefore, such requests should not be made before considering whether the issue can be resolved by using other interpretive tools.

6 Communicating Compliance Information

Communication is an essential and pervasive element of achieving environmental compliance. Not only must organizations communicate within themselves to inform employees about what must be done to ensure compliance, but they must also communicate with government agencies. Failing to communicate adequately with employees can result in failure to achieve compliance, while failing to communicate adequately with the government can result in misunderstandings and enforcement liability. In the worst cases, poor communication can become a basis for criminal liability. Although communicating takes effort, environmental professionals can greatly improve their skills by focusing on the purpose behind the various types of communication and becoming sensitive to the compliance issues such communications entail.

6.1 COMMUNICATING WITHIN AN ORGANIZATION

Environmental compliance involves the application of complex and detailed laws to a multitude of activities. The regulated activities often involve several layers of employees and may be regulated by a multitude of regulatory programs. Compliance requirements that apply to some activities may also affect other activities, such as when production scheduling needs to be coordinated with sales to ensure that effluent limits are not exceeded. Ensuring compliance thus requires organizations to inform their employees about the requirements that apply to their activities and to coordinate functions that are regulated by environmental laws. Environmental professionals should remember a cardinal rule to ensure that complex requirements are communicated in such a way that organizations can maximize the ability to achieve total environmental compliance: *Simplicity is a virtue.*

6.1.1 Reducing Uncertainty About Requirements and Responsibilities

Properly executed efforts to communicate compliance information will provide employees with the information they need in an efficient format. Depending on the circumstances, compliance information can be communicated orally or in writing. In many instances, it may be advisable to do both. For example, employees can attend a training session that covers compliance issues and can also be provided with written materials they can refer to after training, but the dissemination of compliance information cannot be limited to training. Often, it will be necessary to provide employees with information through such means as posting signs and making response plans and similar compliance materials available. Regardless of how information is provided, successful dissemination usually hinges on providing the important information in clear language and communicating what is expected of employees.

Environmental requirements will be difficult to comprehend unless the organization makes the effort to express them clearly to employees. The objective of such efforts is to ensure that employees know the requirements the organization must meet and understand their personal responsibilities in meeting those requirements. Unfortunately, clear communication has not always been the hallmark of the environmental professions. Although many environmental professionals are good communicators, the profession as a whole is inclined to adopt practices that impair the communication of compliance information.

As an initial step, environmental professionals should avoid practices that impair comprehension of the information they are trying to communicate. The propensity to use regulatory jargon and acronyms, combined with presentation skills that tend to be oriented toward technical audiences, can confuse people who do not have environmental backgrounds. It is especially easy to lose an audience that does not understand an acronym or citation. For example, advice that "a 5(a)(1) can be avoided by preparing and implementing a RAP" may not receive serious attention from managers who have no idea about what is being discussed. Unfortunately, while some environmental professionals feel that the prolific use of acronyms and regulatory citations conveys an aura of competence, this practice can convey that the person is pretentious or even foppish. Furthermore, few people object when professionals use common and readily understood words to express complex or technical subjects.

Besides avoiding excessive use of jargon and acronyms, environmental professionals can enhance the effectiveness of communications by making them relevant to the audience. A common fault in preparing materials that are intended for the use of employees is to draft them in a style that evinces concern for the environment and commitment to compliance but falls short in providing substantive guidance on how to comply with environmental requirements. For example, procedures and plans are often replete with platitudes, abstract directives, and paraphrased regulatory text and devoid of information that specifically applies to the facility. Statements in internal procedures such as "Regurgitron Corporation treasures the environment and will spare no effort to prevent stormwater pollution" may sound great (or perhaps insincere) to the government and the public but does nothing to help employees comply with environmental requirements. Unless an organization's environmental documents provide employees with information on how to carry out their responsibilities, their inspirational and philosophic content will do little to enhance compliance.

One aspect of achieving relevance is to make communications focus on what employees need to know. For example, forklift operators at a silicon wafer plant are less likely to benefit from a video showing how airports can prevent stormwater pollution than they will from a live session where the trainer tells them how to recognize and prevent stormwater contamination at their facility. In addition, the content of materials used to promote compliance should be limited to the issue at hand. For example, it is not uncommon for oil spill prevention, control, and countermeasure plans to contain information that is neither legally required nor particularly relevant to preventing and responding to oil spills. Including unnecessary material in such documents makes it more difficult for employees to find the time to read the document and burdens them with the task of assessing what material is relevant. In general, compliance documents work best when they focus on the issue at hand and do not attempt to serve as environmental treatises.

Specificity is a critical element of making environmental documents relevant to employees. Unfortunately, a lack of specificity is commonly seen in documents that purport to explain to employees what needs to be done to comply with environmental laws. For instance, a document that states that stormwater outfalls must be "inspected periodically" is unlikely to encourage compliance with a stormwater permit, as the lack of specificity does not impose a requirement for action on any particular employee nor does it tell them what they are supposed to do. Relevant guidance will provide employees with information such as when stormwater inspections should be done, who should do them, what they should be looking for, and what to do if pollution is observed. Compliance guidance should be drafted in ways that avoid generalizing employee duties, which forces employees to interpret what they are being told to do. Such generalized guidance can often be identified by a document's reliance on general adjectives and adverbs such as "periodic," "as necessary," and "promptly" to communicate procedures. Because general descriptors require employees to guess at their meaning, they increase the potential for noncompliance by increasing the likelihood that employees will make inappropriate interpretations. Furthermore, unless organizations provide employees with guidance that is sufficiently explicit to inform them about how to carry out their duties, they risk delegating the responsibility of exercising critical judgment to the employees least qualified to make those decisions.

To maximize effectiveness, the communication of compliance information should put the desired action in tangible contexts. Procedures are often more effective when tailored to specific activities. For example, a general corporate procedure that tells employees to use sorbents to clean up minor leaks of oil and to dispose of them properly will not have the same effect as facility-specific procedures that describe the type of sorbent to be used, where the sorbents are stored, and where sorbents should be disposed when spent. Contextual guidance not only tends to provide more specific information than general guidance, but it is also more credible. Compliance materials should also address how employees should respond to environmental aspects during all the situations they are likely to encounter. For example, it is not enough to describe routine duties if compliance depends on employee action during upsets and maintenance shutdowns.

Training activities should also make allowances for the fact that employees are frequently unfamiliar with environmental regulatory schemes and, if not educated by their employers, will tend to guess at what is required. This is particularly problematic when legal requirements do not have an obvious relationship to environmental protection or seem overly prescriptive. Unless common areas of misunderstandings and the tendency to rationalize noncompliance are addressed in training and guidance materials, environmental compliance can be jeopardized in critical ways. For example, employees commonly misunderstand that *de minimis* emissions and discharges that appear to have little effect on the environment can nonetheless be prohibited by law.

Employees also frequently misunderstand the legal effect (or lack thereof) of government assurances and impending regulatory actions such as pending permits and proposed regulations. Although employees may feel justified in proceeding with certain activities on the basis that an agency is aware of the conduct or because an application for a required permit has been submitted, such actions do not authorize noncompliant conduct. To avoid noncompliance arising from such misconceptions, they should be addressed as part of compliance training.

Organizations should also provide their employees with guidance as to what to do in the event of problems such as control equipment upsets and sampling errors. Such events will almost always require the exercise of management judgment, and employees should be instructed to notify management promptly when they occur. If not instructed to do so, employees may decide to exercise their own judgment which may result in serious compliance problems.

6.1.2 Preparing Environmental Plans

Environmental plans form key elements in many regulatory programs and are frequently required by regulations and permits. One of the notable features of these plans is that they are prepared by the regulated party instead of a regulatory agency. Typical issues addressed by plans are emergency response, best management practices, monitoring, and inspections. Although environmental plans are tailored to individual facilities, they constitute enforceable requirements, just as regulations and permits do. This means that regulated parties are subject to penalties if their plans fail to meet regulatory requirements or are inadequately implemented.

There are two reasons why organizations normally prepare environmental plans for their facilities. The first is that plans are required by either a regulation or a permit condition. The second is that the plan is perceived as an effective tool for informing employees what they need do to comply with environmental requirements. Although the plethora of plans required by regulations may cast doubt on whether employees have the time or ability to read and comprehend all their requirements, their preparation is a legal requirement, and organizations should not allow doubts about the effectiveness of these plans as environmental tools dissuade them from following the often detailed requirements for their preparation. Furthermore, plans can provide employees with crucial information for situations where concise guidance is needed under exigent circumstances.

The first step in preparing an environmental plan required by a regulation or permit is to read and carefully follow the requirements for the plan. Although it may seem simple to ensure that plans contain all the required elements, many preparers fail to read carefully the provisions governing content and thus draft plans with significant gaps. The next step in preparing an environmental plan is to ensure that they are written for the proper audience. Many plans seem to be aimed more at impressing regulatory agencies than helping employees achieve compliance. Such plans tend to be permeated with statements that applaud the good intentions of the organization yet provide only minimal or unclear guidance to the employees who must implement the plan. For example, plans that consist mostly of statements such as "Slop Corporation will respond to any releases of oil with timely and appropriate efforts to prevent serious environmental damage in full accordance with any appropriate international, federal, state, and local laws" may sound superficially appealing but rarely tell a mechanic what to do when he or she spills a quart of motor oil while servicing a vehicle. A better plan might describe the locations of spill kits and provide instructions on how to apply and dispose of sorbents.

The scope of an environmental plan needs to be thoroughly considered. There is a tendency for plans to deal with the worst-case scenarios yet omit guidance for more mundane incidents. For example, it is common to see oil spill response plans that deal with catastrophic tank failures but overlook the more common incidents that result in minor drippage or small spills. Another common problem, especially in plans prepared by consultants, is incorporating superfluous materials into plans. Sometimes information not required by a regulation or permit is genuinely useful in a plan, such as when it helps an employee comply with legal obligations such as reporting releases of oil. More often, superfluous materials provide information of dubious value and impede timely comprehension of the plan. For example, spill response plans that provide information irrelevant to preventing or controlling releases become more time consuming to read and more difficult to comprehend. In addition, the readability of most plans can be improved by omitting unnecessary background material.

Superfluous information can also increase the possibility of enforcement by creating enforceable requirements where none existed before. By incorporating burdensome procedures into environmental plans, organizations increase the possibility that the plan will not be executed fully as required by the regulation or permit. Because the failure to follow procedures set forth in plans can constitute a violation, organizations that incorporate unnecessary requirements into plans may be liable for violating requirements that they alone created.

The format and style of plans are important considerations in promoting readability and substantive content. Some employees are slow readers and unskilled at flipping back and forth from the main text to appendices as a means of obtaining information. To be effective, plans should be concise, focused, and explicit. In general, plans should avoid the overuse of acronyms, irrelevant material, and appendices. Specificity is critical. General statements, such as "Regulatory reporting of oil spills is of critical importance and the proper authorities should be notified within allowable time frames whenever appropriate or mandated by existing legal requirements," may be easy to write but are woefully inadequate at telling employees how to comply with legal requirements. It is much better to describe such procedures explicitly — for example, "If an employee observes release of oil from the plant that causes a visible sheen or emulsion in the Waterway, he or she must report it immediately to the plant manager or superintendent who shall immediately call the National Response Center at (800) 424-8802 and provide the information requested."[35]

Specificity usually demands that plans be tailored to individual facilities. Some organizations attempt to save cost and effort by using generic plans that are intended to apply to classes of facilities. Even when these plans contain appendices that provide information specific to individual facilities, they nonetheless tend to be inaccurate and overlook important aspects at individual facilities. Furthermore, this approach to drafting plans tends to result in plans that provide little substantive information and are rarely updated.

Finally, it must be emphasized that the preparation of a plan in and of itself will do little to ensure environmental compliance. Once written, an environmental plan must be properly implemented. This means that the plan must be read by the persons who will be required to carry out its procedures and that employees must be adequately trained on its contents. Plans should also be kept in locations where they are available when needed. They also need to be amended as circumstances change or when inadequacies are discovered.

6.2 COMMUNICATING WITH GOVERNMENT AGENCIES

Communication with agencies that administer regulatory programs is unavoidable given the pervasive regulation of activities that affect the environment. Environmental laws require organizations to communicate with agencies in many ways. Examples include submission of permit applications, reporting releases to the environment, and maintenance of records for government agencies to peruse at their discretion. In addition, communication is generally required in other settings such as compliance inspections. Communication also can present compliance risks, as negligent and willful acts of miscommunication can result in enforcement. Willful acts such as reporting false monitoring data constitute criminal acts and are subject to intense enforcement; however, even unintentional omissions and mistakes in providing information to government agencies can violate strict liability provisions in regulations and permits and thus lead to civil or administrative enforcement. Communication, therefore, should be treated as a compliance obligation and given serious attention.

Because accurate and complete communication by regulated parties is essential if regulatory agencies are to implement their programs successfully, communication issues are extremely important to them. This is one of the reasons why agencies usually treat false reporting as criminal offenses, but even negligence in reporting can be enforced through civil and administrative processes. To ensure compliance with communication requirements, environmental professionals should understand the basic legal requirements concerning communication with the government and should adopt practices that minimize the possibility of erroneous and even ambiguous communications.

6.2.1 Liability for False and Incorrect Communications

Liability for incorrect statements made to government agencies can be classified according to whether the statements were made deliberately or unintentionally. Because most environmental laws subject regulated parties to the legal standard known as *strict liability*, it is only necessary for the statement to be wrong for a party to be found liable. The fact that a mistake was made in good faith or that the party was unaware that information was required to be submitted to the government is irrelevant to the issue of liability. Persons who make statements they know to be false can be held criminally liable.[60,61] As a general rule, liability under these laws requires that the person responsible for the statement knew that it was false and but does not require a finding that the person had a deceptive motive. In addition, parties are not criminally liable for statements that are inaccurate but not known to be false. Therefore, common mistakes and information submitted in good faith but later found to be wrong do not present the risk of criminal prosecution, but such statements can subject employers to civil and administrative enforcement.[62]

Because few environmental professionals intentionally attempt to mislead government officials, the legal issues associated with inaccurate information are discussed first. The most serious risk associated with providing inaccurate or ambiguous information to the government is that it can be perceived as an attempt at deception. This is particularly true when the agency may act on the information differently that it would if accurate or explicit facts were provided. Such communications are particularly problematic, as it can be very difficult to discount allegations that the inaccuracies or ambiguities were motivated by a desire to avoid regulatory oversight or to conceal noncompliance. For instance, inaccurate representations about the amounts of air pollutants that will be released from a proposed facility can mislead agencies into determinations that avert the installation of expensive control equipment. Although estimating emissions is difficult and necessarily imprecise, it can be almost impossible to persuade agencies that inaccurate estimates were developed with innocent intentions when the result enables an organization to avoid spending millions of dollars on pollution control equipment.

Although unintentionally false or ambiguous statements that are perceived as deceptive can increase the risk of enforcement, intentional misstatements or omissions of important information carry a much stronger possibility of criminal enforcement against the organization and the individuals responsible for the deception. One of the broadest sources of liability in the environmental area are federal and state false statement statutes prohibiting the provision of false information to the government.[60,61] These are criminal laws that make the knowing submission of material false information to government agencies punishable by high fines and incarceration. The purpose behind false statement statutes is to protect government agencies from deceptive practices that have the potential to impede or frustrate their governmental functions. The federal false statement statute covers a broad range of activities affecting the federal government, including statements made by parties operating sources that are regulated under the Clean Air Act and Clean Water Act. In addition, environmental statutes such as the Clean Air Act and Clean Water Act specifically make it a crime to provide false information regarding those programs to government agencies.[63,64]

To sustain a conviction under the federal false statement statute, the government must prove that the defendant voluntarily and intentionally made a misrepresentation with regard to a matter within the jurisdiction of a federal agency. Incorrect statements made because of mistake, accident, or other innocent reasons do not fall within the coverage of the statute, but this statute has broad application to environmental matters, as it applies to statements made to state agencies that are administering federal programs such as those under the Clean Air Act and Clean Water Act. This means that the federal government is able in many cases to prosecute for false statements made to state agencies. Furthermore, the government does not have to prove that a defendant knew that a matter was within the jurisdiction of a federal agency. For example, a manager of a large manufacturing plant can be held liable for falsely certifying compliance with the terms of a state-issued stormwater permit regardless of whether the manager knew that the state was administering the federal program.

False statement liability can apply regardless of whether the information was required by law to be provided to agencies or was offered voluntarily; however, withholding information from a government agency is punishable under the false statement statute only if there is an affirmative legal duty to supply the information. In the absence of a requirement to provide information, parties need not inform the government even if the matter has a potentially serious effect on the environment or the administration of a regulatory program. For example, parties are not required to inform the government when they discover noncompliance if there is no regulation or permit condition that requires such notifications. There are many situations, though, in which regulated parties must provide information to the government. Common examples include requirements to certify compliance with permits, report upsets and permit violations, and to provide immediate notifications about releases of oil and hazardous substances. Failure to provide such information when required to do so by law can thus result in prosecution.

6.2.2 General Guidelines for Communicating with the Government

Because inaccurate and ambiguous information has the potential to create serious compliance problems, regulated parties should strive to make statements to agencies as clear and accurate as possible under the circumstances. As discussed previously, providing inaccurate information to the government can be a basis for liability, but it can also affect environmental compliance in other ways. In some cases, providing inaccurate or incomplete information may act to the direct detriment of the regulated party, such as when agencies use the information for such purposes as preparing permits. For example, the failure to describe an outfall in a permit application could result in the outfall not being covered by the permit. Furthermore, errors such as these may not be readily correctable, as agencies are often reluctant to modify permit provisions when the problem could have been avoided by being more careful in providing the required information.

Regardless of whether inaccurate communications constitute violations or aggravate existing burdens, environmental professionals should avoid them whenever possible. Most problems associated with communications that involve compliance matters result from incompleteness or ambiguity. By making conscious efforts to ensure that communications are complete and explicit, environmental professionals can not only avoid misunderstandings but also establish reputations as competent and trustworthy individuals. On the other hand, a reputation for misleading and sloppy communications will diminish a professional's credibility and possibly expose employers and clients to enforcement activity.

Completeness in many cases must be assessed against the regulatory requirement for the communication. For example, permit applications are often required by regulation to provide specific information and to provide it in specific ways. Unless the person making the communication is familiar with these requirements, it is difficult to ensure that the communication will be complete. The best way to ensure that requirements for information are completely satisfied is to review them before the communication is prepared and to review them again before the communication is made or delivered.

Another aspect of completeness is that all the relevant information should be reported. Failing to report all the relevant information can damage credibility and result in enhanced regulatory scrutiny. In addition, it is important to explain significant points that may not be immediately obvious to a regulatory agency. For example, if a spill has released hazardous substances into a school yard or residential area, this information should be reported to the government, even if the reporting regulations technically do not require that this information be provided.

Clarity and objectivity are important if misunderstandings are to be avoided. Whenever possible, provide information precisely so that agencies do not draw inappropriate conclusions. Sometimes it will be impossible to provide precise information to a government agency because precise knowledge is impossible under the circumstances. For instance, an organization may not be able to ascertain the quantities of hazardous substances released after a spill because there is no way to measure the amount precisely. In such situations, it is critical not to make unfounded guesses when providing information. When the circumstances make it impossible to know the facts with certainty, organizations can reduce their potential liability by disclosing that their information may be inaccurate or unreliable and reporting what they do know. However, when better information becomes available, an organization should take care to see that the agency is updated.

Environmental professionals should be careful to avoid terms that can readily be misconstrued or misinterpreted. People often perceive the meaning of many common adjectives and adverbs differently, and using such words in regulatory communications invites misunderstandings. Specific words that should be avoided whenever possible are "immediately," "empty," and "small," as they are subject to variable interpretation. Examples of better descriptors, depending on the circumstances, might be "within ten minutes," "no visible contents," or "less than one pound."

Another important aspect to providing information to agencies is to avoid superfluous or irrelevant information. This does not mean that regulated parties should withhold information that agencies may find useful in addressing environmental matters. Government agencies, however, are generally inappropriate forums for attacking individual employees or expressing discontent with company policy; such statements make organizations look bad and reflect poorly on the individuals who make them. Although some degree of judgment is required when determining what should be disclosed to the government, some areas are clearly the province of the regulated party and should be addressed internally.

Providing superfluous information can also tax attention spans and increase the possibility that misunderstandings will occur. This can be particularly important in oral communications where no records are kept. If a recitation of superfluous or irrelevant information prevents a regulator from comprehending the crux of a communication, it may be nearly impossible to avoid compliance problems if the communication was intended to address a compliance requirement. For example, if a notification by telephone about an upset was misunderstood because the caller spent most of the conversation talking about proposed improvements to the treatment system, a serious dispute can arise should the agency contact maintain that the upset issue was not sufficiently raised to preserve the defense.

6.2.3 Compliance Certifications

Environmental regulations and permits are increasingly incorporating requirements that organizations certify that facilities are fully complying with environmental laws. For example, certification provisions are found in many air emission permits, stormwater permits, hazardous waste disposal regulations, and government contracts. These provisions require senior managers such as corporate officers, division managers, and plant managers to sign and submit statements that a facility for which they are responsible is in compliance with the environmental laws applicable to the certification. The major legal consequence of a certification requirement is that evidence is created that connects senior management to the environmental conduct of their organizations. By requiring senior managers to sign compliance certifications, the government forces them to participate in the compliance function in a way that makes them directly liable if they sign a certification they know or should know is false. Such evidence makes it easier for the government to present a case for management complicity in noncompliance; therefore, certifications should be treated as serious compliance matters.

Regulations that require compliance certifications impose the obligation that the signatory manager make a reasonable inquiry regarding the compliance status of the relevant facility and truthfully attest that the facility is in compliance. Unfortunately, government agencies rarely provide clear guidance as to what constitutes a reasonable inquiry and leave organizations guessing as to what is expected of them. Organizations, though, do much to protect themselves and their signatory managers when they apply a deliberate approach to compliance certifications. The first step should be to ascertain who is the best person to make the certification, as the failure to meet the requirements for signatories constitutes a violation. Generally, the class of eligible signatories will be defined by a regulation, the terms of the permit, or an information request. Common provisions require that the signatory be an officer of a company or an equivalent official of a government body. Some provisions also allow organizations to delegate signature authority to such persons as managers of large facilities. Following this, the organization should ascertain which compliance requirements are covered by the certification provision. The next step is to make a reasonable inquiry. In many cases, all that will be required is to check with persons who are knowledgeable about matters within the scope of certification. Some certifications, such as those associated with information requests, may warrant more substantial consultations to verify that knowledgeable persons actually consulted the records needed to ensure that the inquiry was reasonable. It can also be a good practice for managers to protect themselves by noting which persons were consulted prior to signing a certification.

Although signatory managers cannot be liable for incorrect certifications if they make a diligent and reasonable certification and do not know about instances of noncompliance, they must also be able to demonstrate that any errors were made in good faith if they are to reduce the chances of being accused of false certifications. For instance, deliberately shielding oneself from the knowledge of noncompliance will not satisfy the requirement to engage in reasonable inquiry.[27] Likewise, relying on assurances from seemingly dependable subordinates may not hold up when scrutinized under the harsh light of hindsight.[28] For example, the good faith of a certification may be questioned if it appears that the subordinate employees did not actually make the effort needed to ascertain compliance.

Such difficulties are best avoided by adopting systems and practices intended to ensure total environmental compliance. By establishing a secure foundation in this respect, organizations and their managers can avoid many of the problematic issues associated with compliance certifications. In addition to compliance efforts, organizations need to implement procedures that describe what employees need to do when providing information to persons who will be signing compliance certifications. These procedures should set forth obligations to notify key persons of issues that involve potential noncompliance and define the scope of what subordinate employees need to evaluate when a certification is imminent.

In the event that noncompliance is discovered or suspected, it is important that compliance not be certified, as such an action would constitute a false statement. Although few agencies specifically address in permits or regulations what parties must do when compliance cannot be certified in good faith, informal guidance can be obtained from the agency administering the certification requirement. It is also possible to respond to these situations by signing a qualified certification that acknowledges either known or potential noncompliance. Unfortunately, this approach has the potential of constituting an independent violation and also exposes the organization to enforcement for the disclosed noncompliance. Furthermore, the privilege against self-incrimination will probably not preclude agencies from requiring certifications of compliance from regulated parties, as the government's position is that such certifications serve an essentially regulatory purpose rather than a law enforcement purpose to which full constitutional protections apply. Nonetheless, a qualified certification disclosing some noncompliance should present a lesser degree of risk than making certifications that are known to be false.

6.2.4 Facility Inspections

Inspections of facilities by government agencies commonly reveal noncompliance, and organizations should know how to conduct themselves when confronted with inspections. The first step in preparing for government inspections is to ensure that the facility is meeting its environmental requirements. Because many inspections are unannounced, the facility must be in compliance all the time. The purpose of most government inspections is to obtain information about compliance at the facility, and the inspectors will typically want to view the facilities and any documents that are required by the environmental laws encompassed by the inspection. Facilities that present an orderly appearance and maintain well-organized documents are more likely to be perceived as responsible, which can provide the assurances inspectors need to terminate visits promptly.

Even when facilities are in complete compliance with environmental laws, it should be noted that inspections are not without risk. Government inspectors sometimes perceive compliance problems because they misunderstand the legal requirements or apply an overzealous interpretation. Organizations can give the impression of noncompliance if they provide incorrect information to inspectors. To minimize these risks, organizations should prepare for government inspections to ensure they create a favorable impression.

One issue that organizations should decide before an inspection is how forthcoming they want to be during inspections. Most inspections are conducted in a cordial manner but can result in enforcement actions if violations are observed. Although cordiality should be maintained, organizations are not obligated to disclose information that is not requested during inspections. The potential detrimental effect of volunteering compliance information is that it can indicate to inspectors that some additional areas should be evaluated for noncompliance which can result in longer inspections. The potential benefit of providing information that is not requested is that it can create an impression of competence and foster good will. Naturally, an important factor that organizations will want to consider in deciding how open to be during inspections is their level of compliance. In this respect, organizations that have achieved total environmental compliance have far more flexibility in determining how they interact with agencies.

No matter what an organization's practice is regarding how much information should be provided during government inspections, they should ensure that the information they do provide is truthful and accurate. Furthermore, deliberate concealment of information should be absolutely avoided. Providing false information or hiding compliance problems not only creates a negative impression if discovered but also provides separate and sometimes more serious grounds for enforcement.

To avoid confusion when confronted with an unannounced inspection, organizations should designate in advance the persons who are authorized to accompany inspectors during their visits. Such employees should be familiar with the facility's operations and applicable environmental laws and should know where compliance materials such as records and plans are located. They also must be skilled at dealing with people respectfully and understand the need for honest communication. Because organizations will not always be able to have the primary environmental contact available at inspections, several employees at each facility should be trained on how to conduct themselves during inspections.

Most environmental laws and permits require that organizations give the government reasonable access to regulated facilities. In general, organizations cannot ban access for reasons such as safety concerns nor can they require government inspectors to sign documents that release the organization from liability for injuries or otherwise condition entry into the facility. Nonetheless, government inspectors should be advised of specific hazards and offered appropriate protective equipment, if available. Most inspectors will honor *bona fide* requests not to enter hazardous areas such as asbestos abatement enclosures, provided there are alternative ways to gather information. Similarly, requests not to disturb sensitive operations such as activities in clean rooms are usually honored. Should inspectors insist on exposing themselves to unreasonable danger or causing substantial disruptions to operations, it should be suggested that the inspectors confer with their supervisors before doing so. If such suggestions are dismissed, the facility manager may consider contacting the inspectors' supervisors to express concern that their actions are unreasonable under the circumstances.

As a general rule, organizations are entitled to have their representatives accompany inspectors during site visits. Escorts are important both to ensure the safety of the inspectors and to ascertain the scope of the inspection. Prior to inspecting either the operations or documents, the inspector should describe the scope of the inspection to assist the organization in ensuring that it will be able to provide the relevant information. During the inspection, the organization's representative should take notes concerning what was inspected and take photographs of subjects that are photographed by the inspector. Similarly, if the inspectors take samples, the organization should obtain a split and either archive it or have it analyzed for the parameters that will be analyzed by the agency.

Inspectors are usually entitled to review records that pertain to the matter being inspected. For example, an inspector evaluating compliance with stormwater requirements has the authority to review the stormwater pollution prevention plan and monitoring records. Inspectors do not have the authority to review business records not related to environmental matters or documents protected by an evidentiary privilege such as the attorney-client privilege or an environmental audit privilege (although federal agencies generally maintain that they are not bound by state environmental audit privileges). Similarly, it is the policy of many government agencies not to request access to environmental audit reports.[42] In addition, although organizations cannot prohibit an inspector's access to trade secrets, there are measures they can take to ensure that the government does not divulge them to other parties. However, to protect effectively the confidentiality of privileged documents and trade secrets, the persons accompanying the inspectors must know what materials are subject to protection.

At the end of an inspection, most inspectors are willing to hold a closing conference to discuss their initial findings. These conferences are helpful, as they provide an overview of any concerns the inspector may have and give the organization an opportunity to correct any erroneous perceptions. During the conference, it is helpful to ask for an overall assessment of the facility. One way to do this is to ask for a grade or a rating on a scale of 1 to 10. Favorable responses can be useful later should a disagreement over compliance arise. Also, it is important to find out whether the agency plans to take further action as a result of the inspection. If the inspector will prepare a report, ask for a copy. In addition, senior management should be informed about the inspection.

6.2.5 Reporting Environmental Releases

Failing to notify the proper government agencies after a facility has released a reportable quantity of oil or a hazardous substance can result in serious enforcement liability. Many people have been prosecuted criminally for failing to report releases, even though some of them probably had no actual intent to violate reporting requirements. Even people in low-level supervisory positions can be liable for failing to report releases if they are in charge of the facility at the time the release is discovered. For example, the federal government once prosecuted a maintenance foreman for failing to report the "release" of some paint cans that he had instructed his work crew to bury.[65]

Some people are reluctant to report releases because they fear that the reports will expose them or their employers to enforcement actions or expensive clean ups, and others do not report releases because they are unsure of the legal requirements. In any case, failing to make required notifications is a foolish practice. Regardless of whether or not a state or federal agency ultimately determines that action is needed to respond to a release, the agencies take the obligation to report releases very seriously. Ironically, considering the high penalties associated with failing to report releases, government agencies usually do not respond to notifications in ways that result in high costs to the reporting facility. In addition, although some releases constitute violations of law, in most instances the only exposure to penalties will come from failing to report.

The general guidelines for communicating compliance information apply to reporting releases. Employees should make clear that the quantities they report are estimates and that it is better to make a notification when in doubt. They should also avoid using subjective descriptions and instead provide information that is objective and preferably quantitative. When the reporting requirements mandate that several agencies be notified of a release, the organization should make all the required notifications personally, as reporting offices, such as the National Response Center (NRC) operated by the U.S. Coast Guard, will not necessarily relay the reported information to other agencies.

One of the reasons releases present difficult compliance issues is that release reporting laws tend to be complicated. There are several laws that require reporting, although in many cases their requirements are similar. Typically, such laws require facilities to report releases of designated hazardous substances when the amount released in a 24-hour period equals or exceeds a reportable quantity. Reporting laws become especially complicated when applied to mixtures of substances and to frequent or continuous releases; therefore, when a release occurs it can be difficult to determine whether it needs to be reported unless a good source of advice is readily available.

Unfortunately, bad advice regarding whether to report releases is prevalent. Despite the numerous prosecutions of companies and their employees, many people still believe that spill reporting requirements are not taken seriously by the government or that releases are rarely discovered. Other people believe incorrectly that reporting requirements can be ignored when releases present no obvious environmental threat. Bad advice, however, will not prevent a regulated party from becoming the subject of enforcement activity.

Getting good advice on whether and how to report releases can be difficult under certain circumstances. For instance, releases can occur at times when legal and environmental resources are not available to advise facility operators what they need to do. Also, in the absence of prior planning, organizations may fail to comply with reporting requirements when spills or releases occur at inopportune times. One means of promoting compliance is to prepare reporting plans that set forth the procedures to follow when a release occurs. Such plans should contain sufficient guidance to instruct employees on how to determine whether a release exceeds a reportable quantity and which agencies need to be notified. In addition, the plan should provide the appropriate telephone numbers and addresses.

To be effective, such plans need to be oriented to a specific facility and encompass all the substances that can be released at that facility; therefore, the first step in preparing a release reporting plan should be to identify all the substances used at the facility that are present in potentially reportable quantities. This step should include hazardous as well as nonhazardous substances, as one of the first questions that must be answered after a release is whether the substance must be reported. Good sources of information about substances present at a facility are Material Safety Data Sheets (MSDS), vendor records, and plant inventories. The next step is to determine which of these substances are designated by statute as being ones for which releases are to be reported. Among the federal statutes that prescribe reporting requirements are the Clean Water Act; the Comprehensive Environmental Response, Compensation, and Liability Act (CERCLA); and the Emergency Planning and Community Right to Know Act (EPCRA).[66–68] It is also necessary to determine if a substance would constitute a hazardous waste if released, as many Resource Conservation and Recovery Act (RCRA) wastes have reporting requirements separate from those of their constituent substances. For example, a release of slurried lime would not be reportable for its lime content, but the pH could make spilled slurry a hazardous waste that must be reported. In addition, some states have their own reporting requirements.

When a substance has been spilled, it is necessary to determine whether a reportable quantity has been released. Such determinations can be difficult enough when the substance is in pure form but become even more complicated when substances exist as components of mixtures. Under the typical reporting requirement for mixtures for which the quantities of all the hazardous constituents are known, a release must be reported only when a constituent is released in an amount that equals or exceeds the reportable quantity. However, if the quantities of all the hazardous constituents in a mixture are not known, then the government must be notified whenever the total amount of the mixture released exceeds a reportable quantity. This is true even if the average or maximum possible concentrations of a hazardous substance are known but the specific concentrations are not.[69]

To eliminate confusion over reporting requirements, the release reporting plan should contain a table or information sheet for each substance and mixture at the facility that has a reasonable potential of being released to the environment. It is important to be comprehensive, because the release reporting plan may be relied on to make a determination that a particular release need not be reported. The plan should specify the reportable quantity for each substance, and it should state the volume of the substance or mixture associated with the reportable quantity. This is because many people are able to estimate volumes better than they estimate mass, particularly when the release involves liquids. For example, many employees find it easier to assess whether a spill of acetone exceeded 750 gallons than to assess whether it exceeded a reportable quantity of 5000 pounds.

The table should also consider whether the material would be considered an RCRA hazardous waste for which reporting is required. For instance, a spill of dry lime (calcium oxide) would not have to be reported under federal statutes because lime in dry form is neither a designated substance nor a hazardous waste. However, if the pH of slurried or wet lime exceeds 12.5 units and the spilled material is not intended to be reused, the release would have to be reported, because it would constitute a hazardous waste with the characteristic of corrosivity for which the reportable quantity is 100 pounds (which is equivalent to about 8 gallons for a typical lime slurry).

Case Study: ***United States v. Pennwalt Corporation*****, No. CR 88-55T (W.D. Wash. 1989)**

In 1984, the plant manager for a chemical plant in Tacoma, WA, drafted an appropriations request for corporate approval to install bag liners in a pair of tanks used to store sodium chlorate solutions. The intended use of the liners was to prolong the useful life of the tanks, which were suffering from corrosion. At an estimated installed cost of $84,000, the bag liners were far less expensive than replacing the tanks, which would have cost more than $300,000. Less expensive alternatives such as steel liners were rejected because of concerns that the condition of the tanks was too poor to permit welding.

On the night of January 2, 1985, one of the tanks collapsed and released what was later estimated to be more than 100,000 gallons of a solution of sodium chlorate. The plant manager made initial reports to the U.S. Coast Guard and the Washington Department of Ecology (WDOE) that 20,000 gallons of solution were spilled on the ground and that some had flowed into a storm drain that discharges into the Hylebos Waterway. Although the volume estimates were increased and reported the following day, the fact that the solution contained significant amounts of sodium dichromate was not reported. However, through internal discussions at the WDOE, the contact person at the agency learned about the sodium dichromate.

After further investigation, the U.S. Attorney's office indicted the company, the plant manager, and three senior company officials for a negligent discharge and failing to report the release of sodium dichromate. The senior officials were later dismissed as part of a plea bargain in which the company pleaded guilty and paid criminal and civil penalties of about $1 million. The plant manager was able to negotiate a separate plea agreement in which he paid a penalty of $5000.

Discussion

When a major release occurs, preparation is critical, as exercising sound judgment can be difficult under demanding circumstances. Unless some planning occurs prior to a release, the chances of making improper decisions regarding what information should be reported to the government greatly increase. As the company and its plant manager found out, inaccurate reporting combined with an imprudent solution to the deteriorated storage tanks resulted in a criminal prosecution. Despite the fact that the plant manager made a timely report to the WDOE about the spill and the agency had collective knowledge that the solution contained sodium dichromate, the government argued that the Coast Guard was not told about this constituent. Circumstantial evidence indicated that the plant manager knew that the solution contained sodium dichromate and possessed regulations that set forth the reporting requirements. The plant manager maintained that he did not report the release of sodium dichromate because he failed to think of it at the time he made the notification.

As with many factual disputes, it is impossible to assess with absolute certainty whether the plant manager intended to withhold information following the spill; however, it is clear that a thoughtful plan on how to report releases could have saved the plant manager and company from the grief and stress of criminal enforcement. If the reporting procedures had been prescribed before the spill occurred, the company's employees could have reduced the chance of misjudgments during a trying situation. Specifically, reporting procedures that provided relevant information about the specific substances on the premises could have minimized the possibility that employees might overlook minor but toxic constituents present in mixtures and solutions when it came time to report a release.

6.2.6 Responding to Information Requests

Most environmental statutes provide government agencies with the authority to request that regulated parties provide information about activities. Agencies use this authority to collect information for purposes such as developing statistical summaries and assessing whether additional regulation is needed and also to request information for the purpose of evaluating whether to commence enforcement actions. Irrespective of the purpose behind an information request, the failure to respond properly to information requests is a violation of environmental laws, and regulated parties need to ensure that their responses are truthful and complete. Government agencies can and do bring enforcement actions when organizations fail to provide all the information requested by the government. Furthermore, any material statement made to the government that is known to be false can expose parties to criminal liability.[60,61,63,64]

Because regulated parties are under strict legal obligations to comply with information requests, they should not take them lightly. It is not uncommon for information requests to prescribe unreasonable deadlines or make demands that are inappropriately burdensome. When such demands are made, some environmental professionals have been known to provide only part of the requested information in the hope that what is submitted will satisfy the agency. This approach is risky, though, and can give the agency grounds for enforcement. One problem with this approach is that it is difficult to assess what will satisfy the agency. It is not unusual for agencies to know or suspect that specific documents exist but still draft the information request in general terms to ensure that all potentially relevant information is provided. Selective responses that exclude some of the information encompassed by the request can upset agencies and cause them to infer that the party from whom information has been requested is acting in bad faith. To make matters worse, bad faith failures to respond fully to information requests often constitute violations that are very easy to prove and are treated harshly in penalty policies.

Given the seriousness of information requests, organizations can implement a few practices to ensure that they make complete responses. The first step is to notify all employees who may have information applicable to the request and ask them to provide this information to the person or persons who will be preparing the response. Because many organizations have several levels and locations that will have relevant materials, it is essential to communicate the request broadly. It is also important to instruct employees not to destroy or discard materials that may contain pertinent information, even if these materials are subject to routine deletion under a document retention policy.

The most common cause of inadequate responses is lower level employees making cursory searches because they do not understand the importance of making a diligent search. Well-intentioned employees may also decide to exclude documents they believe to be of little relevance in order to save copying expenses or reduce the burden on the persons who are preparing the response. The employees who will be collecting the information should be told that they need to take the request seriously and that they should be overly inclusive when evaluating whether materials are potentially relevant to the request. In addition, all employees responsible for collecting information should prepare a memorandum addressed to the person in charge of preparing the response in which they describe which files were searched and whether pertinent information was found.

Although tight deadlines do not generally excuse full compliance with an information request, agencies are frequently willing to extend deadlines provided there are good reasons for why more time is needed. Common reasons for requesting extensions include that the requested information exists in several locations, key personnel are temporarily unavailable, and more time is needed to copy documents. The granting of a request should be documented in correspondence to or from the agency. If an extension is denied, the party from whom the information has been requested should consider documenting the refusal in correspondence in case an unreasonable deadline results in a deficient response. Sometimes agencies will be more amenable to extending parts of a request if some of the information will be provided by the initial response date.

Another common problem associated with information requests is an unreasonable scope of information requested. Sometimes requests are overly broad because the agency is unaware of the volume of material encompassed by the request. In such cases, agencies often agree to reduce the scope of a request when the circumstances are explained. For instance, an agency may be willing to scale back a request for production records if the volume of information is very large. Similarly, requests sometimes describe several types of documents that provide the same information. A common example would be requests for site maps and property descriptions. Organizations may have this information in various formats such as deeds, reports, and site plans. To reduce the burden of searching for all this information, agencies are sometimes willing to reduce the scope of a request to a single set of maps or descriptions when these documents satisfy the need for information.

Ambiguity is a frequent source of regulatory friction, and information requests are no exception. If there is any doubt as to what is being requested by an agency, it is important to ask for clarification. Any clarification should be documented in writing, preferably by correspondence, as responding to an information request is an enforceable obligation. Some parties are tempted to construe ambiguity in such a way as to reduce the information they will provide; however, this approach risks generating bad will and most likely will cause the agency to serve a second request that is broader in scope. Finally, many information requests will require an officer or senior manager of the organization to certify that the response is complete and accurate. Because false certifications to information requests can expose both the organization and the signatory to enforcement, it is important to consider the guidance presented previously regarding compliance certifications.

6.3 MANAGING ENVIRONMENTAL DOCUMENTS

One hallmark of environmental regulation is the enormous amount of documents that regulated parties are required to manage. Not only must parties keep copies of documents generated by government agencies, but they must also create records and submit information to the government. In addition, the trend toward formalized environmental management systems has caused many organizations to further increase the documentation associated with their environmental activities. Good document practices, however, do not develop on their own, and organizations should pay careful attention to how they create, use, and store information.

The ability to manage environmental documents well is a critical skill that affects environmental compliance in several ways. One effect that documents have on the ability to achieve total environmental compliance is that the burden of creating and retaining documents imposes a substantial obligation on the resources of regulated parties. Although documents are useful for maintaining the information necessary to ensure environmental compliance and provide benefits — such as establishing compliance with requirements and describing procedures for employees who are called to fill in for absent workers — they also require extensive time to prepare and store. Another effect of documents is that they increase the vulnerability to enforcement. Poor document management practices can form the basis for violations and constitute adverse evidence that makes defending enforcement actions almost impossible. Although the only sound approach to minimizing enforcement liabilities is to achieve total environmental compliance, environmental professionals need to be aware that documents are zealously sought during litigation because of the damage they can inflict.

The first step in managing environmental documents should be to focus on establishing guidelines for their creation. Some documents are required by law to be prepared and therefore leave organizations little choice if they are to comply. However, the majority of environmental documents may be prepared at the discretion of the organization, so some consideration should be given as to whether or not they are worth the effort required to prepare them. The least efficient practice that organizations can adopt is to document something merely because it can be documented. Instead, organizations should be able to identify a tangible environmental management benefit for each required documentation practice. If a document does not significantly enhance the performance of the compliance function or takes away time from more valuable activities, it is probably not worth the expense in time and resources necessary to prepare it. One of the problems with requiring too much documentation is the time required by employees to prepare them. For example, it can take an employee several times as much time to write a supervisor a memorandum than to communicate the same information orally. If there is no valid need to maintain a record of the communication and only a few people need the information, oral communication is a less expensive alternative. Furthermore, oral communication allows for more prompt clarification and does not require the use of costly resources such as secretarial time.

Another issue associated with document creation that organizations should address is the use of good judgment by employees in deciding what should be recorded. Environmental documents create a record that reflects directly and indirectly on the compliance status of the facility. Employees whose responsibilities include preparing documents required by environmental laws need to exercise due care to ensure that all the requirements are fulfilled. In general, documents required by law should contain all the information specified by the requirement and little else. There is a tendency on the part of some employees to omit information that they deem to be irrelevant but nonetheless is required by a regulation or permit, but even minor omissions, such as the failure to include the facility's address on a PCB annual document log, can result in enforcement. Other employees may show the tendency to include superfluous information in documents required by regulation. Such information can result in problems if it is founded on conjecture, creates confusion, or is otherwise reported inappropriately. For example, because most generators of solid wastes do not use hazardous waste manifests to ship nonhazardous wastes, using these manifests for nonhazardous waste shipments can cause agencies and waste handlers to question whether the waste is in fact hazardous and cause the generator to expend considerable resources to resolve the resulting confusion.

The issue of employee judgment is particularly important when it comes to ancillary documents such as memoranda regarding compliance issues. Some employees are not comfortable unless they are allowed to draft memoranda or keep records that are intended to absolve themselves of responsibility should another employee make a mistake or fail to carry through with a necessary task. Although it is an organizational matter to determine whether the resources consumed by such memoranda outweigh the benefits, such documentation is a bad practice when the preparer uses poor judgment in drafting the contents. For example, unwarranted speculation about environmental effects or management motives can lead agencies and adverse parties to question the organization's attitude and cause them to infer that serious compliance issues exist. Similarly, unfounded accusations regarding coworkers and managers can lead to discord within the organization. In general, organizations should require their employees to follow proper channels if they have concerns about their supervisors, subordinates, and coworkers and to avoid discussing the nonenvironmental aspects of the organization in environmental documents.

The use of electronic mail (e-mail) can be especially problematic in the environmental context because it encourages indiscretion. The ease of using e-mail creates an air of informality that leads many to incorporate off-the-cuff remarks and loose banter into their messages. Similarly, e-mail lends itself readily to quick responses which sometimes result in employees making replies to messages they later come to regret. Imprudent remarks are easily preserved and can create problems later if messages are encompassed by information requests or discovery in litigation. In addition, e-mail messages tend to exist in multiple locations and can sometimes be resurrected electronically even when deleted by the sender and recipient. Organizations that provide their employees with e-mail access should take care to set and enforce guidelines for its appropriate use.

Managing the circulation of documents is another important part of ensuring compliance. The complex nature of environmental regulation dictates that managers and employees not be overwhelmed with materials to be read and filed. Documents should not be circulated merely for the purpose of showing others that documents have been drafted. Rather, document circulation should be restricted to provide information to those who need it. Determining which employees need which information requires that judgment be exercised, but it must be done if information is to be transferred effectively and efficiently. In many cases, consultation with employees is the best means of assessing who needs particular kinds of information.

Organizations rarely seem to formally assess the costs associated with copying, transmitting, and filing documents, although the results of such efforts can be startling. For example, a set of copies of the monthly discharge monitoring reports generated in a single year at a large industrial facility may consume 12 inches or more of file space, and each individual report may require at least 30 minutes to read and comprehend. Sending and storing copies to 10 managers within an organization can thus consume 10 feet or more of file space annually and up to 60 employee-hours or more if the reports are carefully reviewed and filed by each recipient. In reality, most of these managers will not read the reports. Because most will only be interested in whether the facility is meeting the permit conditions, circulating copies of lengthy discharge monitoring reports is an inefficient means of communicating this information. Furthermore, it is rare for more than one employee at a location to need a copy of a discharge monitoring report. Establishing a central filing location and evaluating who needs to see the actual reports can save organizations substantial amounts of employee time and storage space.

Centralized document storage has other compliance benefits. Organizations that rely on the uncoordinated efforts of individual employees to establish filing systems sometimes find themselves unable to locate important documents when they need them most. In some cases, organizations that do not manage their document storage may not be able to establish whether certain types of documents exist. Furthermore, the document problems become aggravated by personnel changes and new recordkeeping requirements. Unplanned storage can result in compliance difficulties by impairing the ability to respond to information requests and by increasing the difficulty of locating compliance documents. Also, operational and monitoring records are often critical for compliance-related functions such as completing permit applications. Being able to locate such information in a timely manner can help organizations to execute these activities better.

While establishing document management practices is important, such programs have a way of succumbing to entropy if not policed adequately. Large organizations can benefit by hiring full-time document management specialists to ensure that the organization follows good document practices. Smaller entities often rely on the efforts of their office managers to manage their documents. Although a planned, deliberate approach to document management takes effort, it allows organizations to sustain compliance much more efficiently than is possible by relying on the unguided judgment and uncoordinated efforts of individual employees.

7 Designing Facilities and Processes To Enhance Compliance

Some of the most troublesome environmental issues are rooted in failures to consider how the physical and operational constraints imposed by designs will make compliance difficult or impossible. Many constraints are obvious, such as treatment or storage facilities that have been designed without sufficient capacity to handle anticipated loads or volumes. Other constraints are more subtle and can cause unexpected violations at times when little can be done to avoid them. Examples include air emission sources that cannot be readily monitored because of their stack design and chemical treatment systems that cannot be calibrated without the risk of incurring a violation. Poor designs can also increase the pressure on employees to evade environmental requirements. For example, the frustration associated with unreliable pollution control and monitoring equipment may tempt employees to tamper with monitoring systems or surreptitiously bypass control equipment. Similarly, equipment that is dangerous or difficult to maintain may cause employees to defer preventive maintenance in violation of permit conditions. Although these kinds of temptation do not justify unlawful action, prudent organizations will nonetheless consider design issues that will help avoid situations that may spur misconduct.

Another reason for paying attention to compliance issues during the design phase of a project is that organizations have the most opportunity to control the overall costs associated with regulatory requirements at this time. For instance, organizations can evaluate different strategies that can affect regulatory costs such as decisions to build several small facilities or a single large one. To maximize the opportunity to assess and control costs, organizations need to consider both direct costs, such as consulting fees for preparing applications, and consequential costs, such as forgone production while waiting for permits to be issued. By considering all the regulatory costs associated with strategic options, organizations can make better decisions regarding how to design and operate their facilities.

7.1 WORKING WITH DESIGN TEAMS TO ENCOMPASS COMPLIANCE ISSUES

A major reason why organizations fail to encompass compliance issues in the design process is that the persons responsible for developing the actual design are not made aware of the compliance issues or the need to consider them. Unfortunately, it is often difficult for design teams to know all the relevant issues, as environmental requirements vary significantly among facilities. Furthermore, the training and experience of many designers do not provide them with a perfect understanding of the actual or potential environmental requirements that are likely to apply to a particular project. For example, the designers of air pollution control equipment may know that the Clean Water Act regulates discharges to surface waters but may not know that a stormwater permit may require a facility to minimize the entrainment of contaminants such as fly ash into stormwater. Such considerations, when known, can often be addressed in designs at minimal cost yet greatly enhance the ability to meet environmental requirements. Conversely, failing to address compliance issues during the initial design can force facilities to undertake expensive modifications to address compliance issues that could have been foreseen but were not considered.

To ensure that compliance issues are evaluated and addressed fully during the design of facilities and modifications, organizations need to deliberately encompass compliance issues into design processes. Because the environmental staff and the management of the facility are usually in the best position to assess compliance concerns and their effect on the contemplated operations, they should assume responsibility for determining what requirements apply to the facility and communicate these issues to the persons who will prepare the design. Although it may sometimes appear simpler to shift the burden of addressing compliance issues to the design team, such teams are less likely to be aware of the potential compliance issues to the same degree that the management of the facility is. Moreover, it is much more expensive to require the design team to educate itself than to brief the team on the relevant compliance concerns.

Even when the design team is familiar with the compliance issues associated with the type of project, it is nonetheless important for the organization to provide an independent assessment of the potential compliance issues. One reason for this is that environmental requirements tend to be specific to individual facilities. In particular, permit conditions will vary among facilities, and many environmental requirements are applied in ways that are tailored to the characteristics associated with individual facilities. For example, stormwater pollution prevention plans tend to set forth requirements that can be difficult to anticipate without detailed knowledge of the facility. Another reason for providing the design team with an assessment of compliance issues is that designers sometimes take approaches that fall short of what is needed to achieve total environmental compliance. For example, some designers believe that aspects of designs that involve violations that have not been historically enforced do not warrant serious attention. Of course, this attitude is inimical to total environmental compliance and can lead to expensive problems should an agency change its enforcement practices.

Having decided to provide a design team with information regarding the compliance issues it wants to be considered during the project, an organization also needs to ensure that the designers consider the issues diligently. With respect to systems that are designed by consultants specifically for the facility, one means to ensure that compliance issues are carefully considered is to make it a condition of the contract. For example, a contract could require the firm to consider the compliance issues associated with specific permits and regulatory requirements and account for all such requirements in the design. A benefit of dealing with this issue in the contract is that it makes the firm aware that compliance is a serious issue at an early stage where such concerns can be addressed most efficiently in terms of engineering and construction costs. When contracting for prefabricated equipment, the flexibility regarding facility-specific compliance issues may not be as great as for projects where the systems will be constructed onsite. However, the degree to which the equipment meets environmental requirements can be considered when deciding on which equipment to purchase, and manufacturers will usually modify equipment that is manufactured to order when requested by customers. To ensure that manufacturers incorporate special requirements, it is generally necessary to negotiate issues involving feasibility and costs before the contract is signed.

7.2 DESIGNING SYSTEMS TO COMPLY UNDER ALL CONDITIONS

Except for some exemptions, such as an upset provision, facilities must comply with environmental requirements all the time. Although it seems logical to expect that units at facilities would be designed to ensure continual compliance, many designs cannot cope with rare but inevitable situations that were not adequately considered when the unit was designed. Examples include equipment that cannot be maintained or calibrated without violating permit conditions, as is the case with some instrumentation associated with pH control systems. Sometimes the compliance issues associated with such events can be resolved by accounting for them in permits, but it is not unusual to encounter situations where regulatory agencies lack the authority to permit the event, or procedures required to obtain the approval are extremely burdensome. Unless the uncommon operating scenarios are addressed during the design process, the organization will have ignored the best opportunities for feasible solutions.

Insufficient capacity is a common design deficiency.[31] Capacity problems are most frequently encountered when systems are unable to handle extreme flows or shock loadings. Most incapacity problems are rooted in a failure to consider the reasonable scope of conditions that the facility will encounter. Extreme loads typically occur because rare events generate atypically high loadings, such as when large storms produce high stormwater flows, or when coincidental events cause high loadings, such as can happen when maintenance activities result in concurrent drainage of process vessels that contain high levels of pollutants. The magnitude of some extreme events such as stormwater flows can often be estimated using probabilistic analysis of prior monitoring data. The capacity needed to meet loading from coincidental events is best determined by analyzing the possibility of such events and estimating the resultant loads.

Although many processes tend to operate over a fairly narrow range of rates most of the time, they are nonetheless subject to variability from causes such as changes in feedstock quality or extreme loads. Changes such as combusting fuels with higher than normal moisture content can affect the combustion products or increase pollutant loadings; therefore, the effect of process variations needs to be considered when preparing a design. Other factors affecting compliance at the extreme ranges of operations should be taken into account during design. For example, facilities that will be subject to wastewater monitoring need to collect flow information to calculate discharge quantities and must ensure that the flow measuring equipment operates over the entire range of flows that can be anticipated. Although most designs consider the ability to measure higher than normal flows, some do not consider the need for auxiliary flow measurement equipment when a facility discharges abnormally low flows, such as during maintenance shutdowns.

Maintainability is an important compliance issue because the performance of most control systems will deteriorate as maintenance is deferred; therefore, equipment should be designed to facilitate the required level of maintenance. When possible, equipment should be selected that does not require special tools or services for maintenance. Furthermore, the equipment should be designed and installed in ways that facilitate quick repair. Among the issues to consider are worker access, parts availability, and intervals at which routine maintenance will be required. It is often desirable to have the capability to perform preventive maintenance without shutting down the process or unit that is being maintained. By considering the effect that maintenance requirements will have on operations, designers can sometimes incorporate features that greatly reduce the costs that maintenance imposes on production and other activities.

Another issue that should be addressed during design is how maintenance activities will affect compliance. It is not uncommon to encounter situations where equipment either cannot be maintained in a manner that ensures compliance or will impose extreme burdens if compliance is to be achieved. For example, installing wastewater treatment aerators that cannot be serviced without shutting down the treatment system will require an industrial facility with a single aeration basin to take lengthy shutdowns if it is to comply with effluent limits when the aerators need servicing. Issues such as these can be very difficult to resolve after the installation is completed but can usually be avoided altogether when given adequate consideration during the design process.

Back-up equipment is sometimes needed to ensure environmental compliance, regardless of whether it is feasible for a facility to shut down following equipment failure. Back-up equipment may also be necessary when regulations impose contingency-related requirements such as spill control systems. The need for such equipment can often be discerned by analyzing the consequences of events such as power outages, fires, or flooding. Particularly sensitive areas to consider are the effects that emergency events will have on control, monitoring, and safety systems. When possible, fail-safe designs should be used to prevent failure events from causing compliance problems. When such designs are not feasible, warning and alarm systems should be considered to alert employees about potential noncompliance.

Although the installation of back-up equipment will not always be warranted from an economic perspective, the issue should nonetheless be evaluated during design. The initial step in such evaluations should be to determine the consequences of the failure of particular equipment items. For example, if a control device fails, will the facility be able to operate in compliance or will it have to be shut down? If required to shut down, how long will it take to repair the equipment and restart the facility? Economic analysis of shutdown scenarios can provide managers with useful information when deciding whether to install backup equipment. Organizations should also consider whether equipment can be designed in ways that reduce the costs associated with failure. For example, designing a treatment system with parallel chambers may allow for limited operations with lower economic losses than single-chamber or linear designs.

7.3 DESIGNING SYSTEMS TO AVOID COMPLIANCE ISSUES

Compliance issues can impose several types of burdens on regulated facilities. For example, the need for organizations to comply with regulatory requirements can delay construction activities, impose recordkeeping and reporting requirements, increase disposal and treatment costs, and force facilities to curtail operations. However, it is often possible to design systems in ways that minimize the number of regulatory requirements that apply. Although some believe that designing to minimize compliance issues involves evading environmental controls by looking for loopholes, it actually tends to result in more environmentally sensitive facilities by incorporating pollution prevention concepts and evaluating the true costs of designs that have greater potential to adversely affect the environment.

The first step in determining whether compliance issues can be avoided is to determine the legal requirements that may apply to the proposed facility. The basic issues to consider include permitting requirements, recordkeeping and reporting, employee workloads, and the costs associated with managing wastestreams. Once the compliance burdens and their associated costs have been identified, planners and designers can use this information to evaluate whether they can be mitigated or avoided by alternative designs. In some instances, the analysis can lead management to reconsider the performance criteria or contemplated approach to the enterprise in order to capitalize on unanticipated opportunities to increase economic efficiency.

As a practical matter, permitting considerations have the greatest potential to affect the economics associated with a major design. Several environmental statutes and regulations prohibit owners or operators from commencing construction activities until they have obtained a permit. Other regulations impose stringent requirements on new sources that may increase permitting burdens. These permitting requirements are generally triggered by factors such as quantities of air emissions, new construction, and extensive refurbishment. Understanding how these requirements apply to proposed designs can avoid unexpected costs and expedite completion of the project.

One of the better known programs involving construction permits is the Prevention of Significant Deterioration (PSD) program under the Clean Air Act.[70] Under this program, owners and operators of certain air emission sources must obtain permits before they construct or modify their facilities. The application requirements are rigorous and frequently require air quality modeling and an intensive analysis of pollution control options. Because the PSD permitting process can take over two years to complete, organizations generally will try to avoid projects subject to PSD when feasible. PSD permits also tend to contain more extensive requirements than would be imposed absent PSD review, which makes the prospect of PSD review even more unattractive. In many cases, sources can lawfully avoid PSD review by designing facilities so they are not subject to the program.

Before even considering whether to design around PSD requirements, a word of caution is in order. The PSD and other new source review programs are structured such that some parties are tempted to abuse the legal framework in order to evade review. The federal government is aware of this potential and has established an enforcement policy that applies to parties who use bad faith tactics to circumvent review;[71,72] therefore, planning and design decisions that may affect the applicability of new source review should be carefully evaluated. In many instances, the organization should consult with legal counsel before proceeding with projects that could arguably be subject to review.

There is no bright-line guidance as to what constitutes a circumvention, but EPA has set forth several indicators. One aspect that regulatory agencies may consider is whether an organization is attempting to evade review by dividing construction or modification projects into phases that individually are not subject to review. For example, filing of two or more applications for minor source permits within a short time period may indicate that an organization is attempting to circumvent new source review. Such filings do not in themselves provide grounds for enforcement but will subject the organization to more scrutiny. Another set of indicators looks to whether economics of the modifications suggest that they constitute parts of a larger project. Documents such as financing applications, reports to directors or shareholders, or independent analyses of the economic viability of the projects may be considered. Agencies will also look at statements made by company representatives to see if they indicate an intent to circumvent new source review. In any case, organizations that have secured several minor source permits over a short period without new source review may find themselves answering a lot of questions regarding their intentions.

Once an agency determines that a source has circumvented new source review, it may decide to commence enforcement. Typically, the goal of enforcement actions in these circumstances is to assess a financial penalty and require the source to install the pollution control equipment that was avoided by circumventing new source review. Typically, agencies will require the installation of control equipment that is state-of-the art, regardless of whether less expensive equipment would have been required had the new source review been completed before the source was constructed or modified. Because the penalties and injunctive relief associated with these kinds of enforcement actions tend to be expensive, organizations should be careful to ensure that they comply with new source review requirements when applicable.

The new source review programs apply to sources that are capable of emitting pollutants in amounts that exceed regulatory thresholds, so the most obvious means to lawfully avoid the permitting process is to design the project so that it will not emit pollutants above those amounts. One approach is to design facilities with construction capacities that emit pollutants below the regulatory thresholds. Whether this approach is practical will depend on economic factors such as economies of scale and the time delays incurred by the permitting process. Sometimes, factors such as the location of markets may justify the construction of smaller multiple facilities instead of a centrally located but larger facility; however, it is important to be aware that constructing and later expanding a small facility may be construed as unlawful circumvention if the small facility was not economically viable when originally constructed.

Another way to avoid new source review is to incorporate control strategies that reduce pollutants below the regulatory thresholds. By installing high-efficiency control technology or incorporating designs that apply pollution prevention principles, the overall emissions of a proposed facility may fall below the pollutant thresholds which require new source review. This approach is particularly prudent for facilities where new source review is likely to result in the imposition of such control strategies anyway. In such cases, there is little to be gained by going through PSD review and installing the appropriate equipment or process modifications that will allow the organization to go forward with the project without new source review. Organizations can even consider whether investing in pollution control equipment or pollution prevention modifications that exceed normal costs is justified by avoiding the time and expense required to obtain new source review permits.

Another approach commonly used to avoid new source review is to incorporate restrictions on operations in air emission permits to ensure that the source does not operate at levels that produce quantities of air pollutants that would subject it to new source requirements. Although the resulting "synthetic minor" source status can avoid new source review, it does have the effect of subjecting the facility to the limitations and restrictions that are intended to lower emissions below the levels regulated by the new source programs. Furthermore, should a synthetic minor source violate the restrictions on operations, the organization could find itself subject to enforcement for violations of the permit and the applicable new source review program.

Consideration of regulatory burdens during the design process should extend to matters other than permitting. One element that deserves much attention is evaluating how designs can avoid regulatory requirements that are costly to meet and also present risks of noncompliance. For example, EPA regulations require facilities that release hazardous substances at rates above 10,000 pounds per year to estimate and report them each year.[73] Such reports generally require substantial expense and technical effort to prepare. Many organizations have faced enforcement liabilities because they did not follow the regulations carefully. Although not all the costs associated with complying with such regulations can be predicted with certainty, they should be considered whenever possible. For example, one process option may require the release of hazardous substances above the reporting threshold, while another option does not. Instead of selecting a process option based on its capital and operating costs, an organization should also consider the tangible and intangible regulatory costs associated with the options. The tangible costs would include the annual cost of preparing the release reports, while intangible costs would include the risk of enforcement actions.

8 Promoting Compliance Through Careful Permitting

The requirements imposed by environmental permits are similar to statutes and regulations in that they carry the full force of the law and can subject organizations that violate them to enforcement by government agencies and private citizens. They differ, however, from statutes and regulations in how they are developed. Statutes and regulations usually affect many parties, and their development usually involves some degree of lobbying and public participation. In many cases, individual regulated parties need not actively participate in the process, as their interests are adequately protected by trade associations and similar organizations. Although most permitting processes provide for public comment, the applicant will bear the brunt of the effort in obtaining the permit because trade associations rarely involve themselves in matters that do not collectively affect their members. If an organization hopes to obtain good permits, it must rely on its own initiative.

One of the best ways a regulated party can avoid compliance problems is to pay close attention to how it obtains environmental permits. Permits have become a major means by which regulatory agencies impose environmental requirements, even though the process is an inefficient and often cumbersome way of regulating activities. Careful attention on the part of regulated parties will allow the permitting process to proceed faster and with less long-term effort. Organizations should be mindful that the time and effort needed to avoid or resolve issues raised during the permitting process increase exponentially as the process progresses. An item that takes several minutes to complete when preparing the application may take several hours to address during the comment period and several days of intensive effort to resolve in a permit appeal. Limitations of resources at the permitting agency may even preclude this mechanism as an option to address what should have been raised before the permit was issued. Organizations that are adroit at permitting understand that focusing resources at the beginning of the process will often expedite the subsequent stages.

8.1 OVERVIEW OF THE PERMITTING PROCESS

Although variations exist, the general process by which federal and state agencies issue and modify environmental permits has the following steps: (1) application, (2) public comment on the draft permit, (3) issuance of the final permit, (4) appeal process, and (5) modification procedures. It is essential that environmental professionals understand the importance of each stage, as aspects that may seem unimportant at one stage can have enormous implications for others. For example, omitting information in an application can preclude review of important issues in an appeal. Environmental professionals also need to understand the importance of the administrative record to the permitting process. This record consists of all the materials the agency relies on when drafting the permit. These materials include the permit application, correspondence relating to the application, comments on draft permits, internal agency documents, and the hearing record. Should a permit be appealed, the administrative record will define the limits of what issues can and cannot be adjudicated during the appeal. The administrative record can also be critical in interpreting the provisions of a permit.

The permitting process normally begins when the permit applicant prepares and submits the application. Application requirements for environmental permits vary widely among agencies and regulatory programs. Some permits, such as the General National Pollutant Discharge Elimination System (NPDES) Permit for the Discharge of Stormwater Associated with Industrial Activity that is issued by the U.S. Environmental Protection Agency (EPA), require the applicant to submit a simple Notice of Intent (NOI) form that identifies the facility and is properly signed. Other applications are extremely complex. For example, applicants for the permits required by the Prevention of Significant Deterioration (PSD) provisions of the Clean Air Act must often collect meteorological data, monitor for ambient levels of pollutants, estimate air emissions, model predicted concentrations of pollutants, and evaluate the technical and economic feasibility of potential control technology.[74] In all cases, however, permit applications are legal documents and deserve serious attention to their requirements and content.

Agencies typically set forth their application requirements in regulations and frequently supply forms to be used in completing the applications. Occasionally, agencies will publish official guidance to assist the applicant in preparing the application. Once completed, applications are submitted to the permitting agency, which uses the information to prepare a draft permit. Applications are important components of the administrative record, as they indicate the scope of what the applicant is attempting to have covered by the permit and reflect much of the information that the agency considered when preparing the permit.

Once written, draft permits are provided to the applicant and made available to the public for review. The review process is intended to allow the applicant and any other interested party to submit comments on the draft permit. In some cases, the agency will hold a public hearing on the permit, but written submissions usually provide the most important information. Once the comment period expires, the agency will assemble the comments, consider them, and prepare a summary of its responses. If the agency feels that a comment justifies a revision to the draft permit, the agency will incorporate a change into the final permit and explain the reason for doing so; otherwise, the agency will briefly explain why it chose not to act on particular comments. Note that the draft permit, comments, and responses to the comments will be incorporated into the administrative record.

Once the final permit is issued, the permittee and the public will have a limited period of time in which to file an appeal should they choose to do so. Most permit appeals are held by the agency that issued the permit and are followed by an opportunity to appeal the agency decision to a court. The typical appeal commences when the permit applicant files a petition specifying the conditions that the applicant is contesting and the reasons for the appeal. The matter will be assigned to an administrative law judge who will conduct a hearing on whether the agency had a legal basis for imposing the contested conditions. As a general rule, the permitting agency will prevail if it can establish that it had any rational basis to impose a condition and followed its procedural rules in doing so. This is a relatively easy standard for agencies to meet and negates the need for the agency to rebut that more reasonable alternatives to the contested conditions were available.

After the administrative process is completed, parties usually have the opportunity to appeal to a court; however, courts tend to apply the same standards of review as administrative tribunals and usually defer to the agency's factual and policy determinations. Moreover, courts will limit their review to matters that appear in the administrative record and refuse to address issues that were not considered during the agency hearing. Although it is technically possible to obtain redress on permitting issues from courts, as a practical matter it is difficult to prevail on matters that were not resolved at the agency level.

8.2 BASIC OBJECTIVES IN OBTAINING ENVIRONMENTAL PERMITS

Too many environmental professionals approach permitting as an administrative requirement that must be fulfilled before their employers or clients may commence regulated activities, but lackadaisical attitudes towards permitting invite compliance problems and increase the vulnerability to enforcement. Unless environmental professionals learn how to handle permitting matters competently and diligently, they will be unable to handle one of the most important compliance functions. The prerequisites to acquiring permitting skills include gaining an appreciation for the role of permitting in the compliance function and developing an understanding of what should be accomplished when obtaining a permit.

Conscious consideration should be given at all stages of the permitting process to the characteristics that the applicant wants to see in the permit. It is essential that the permit cover all the operations that need to be permitted and desirable that the permit allow the organization to modify its operations to accommodate reasonable changes to activities. The permit should also clearly communicate what conduct is expected of the organization and should be readable by the employees who must ensure compliance. Another characteristic of a good permit is that it will reflect technical reality and mandate only requirements that are feasible to meet.

If compliance problems are to be avoided, it is important to visualize how a party will be affected by a permit long before the application is submitted. For example, a facility that discharges wastewater into a river segment that does not always meet water quality standards should recognize and address the potential difficulties of the situation before the government requires such action. Even intermittent instances during which water quality standards are not achieved will likely result in the river segment becoming subject to special regulation. The possible ramifications to the discharger could be discharge restrictions based on mixing zones, pressures to connect to a publicly owned treatment works, or more stringent effluent limitations. By considering such issues ahead of time, organizations can act in ways that minimize their long-term environmental expenditures and avoid confrontations over environmental issues.

Foresight also requires organizations to consider all the important aspects of the permitting process and how the final product will affect the regulated activities. Most issues related to permitting will be of either a technical or legal nature. Unfortunately, many parties make the mistake of focusing on one and neglecting the other. The technical issues tend to get the most attention because they relate to the more immediate concerns such as emission and discharge limits, but organizations must address the legal aspects of permitting if they are to ensure consistent compliance. An example of a legal issue commonly found in permits is the provision setting forth procedures for preserving the upset defenses found in many wastewater discharge permits. Parties that fail to meet the legal requirements can find themselves embroiled in compliance issues that bear little relationship to environmental quality. Nonetheless, it is not unusual for parties to neglect important issues during the permitting process because they have focused on the substantive environmental aspects of the permit and failed to consider administrative ones.

8.3 COMPLETING PERMIT APPLICATIONS

Permit applications are important because they provide information to the permit writer, constitute an important part of the administrative record, and form a basis for interpreting the scope of the final permit. Although the informational purpose of a permit application is obvious, it is surprising how many applications do not communicate all the information necessary to ensure that the permit writer understands what activities are to be covered by the permit. Similarly, many permit applications are completed with little or no consideration to establishing an administrative record. By preparing applications properly, organizations can minimize misunderstandings on the part of the permit writer and ensure full coverage of the activities for which the permit is sought.

Once an organization recognizes that it will need a permit, it should determine what information will be needed to complete the application. Some applications require minimal information, but others can require information that can take over a year to collect. In addition, some applications require the applicant to provide detailed engineering studies which can take months to years to conduct. When substantial effort is required to collect information or to prepare engineering studies, the applicant should consult with the permitting agency to ensure that the contemplated work will satisfy the agency's requirements. Such consultations can also expedite the process by avoiding delays associated with requests from the permit writer for more information. In some cases, permitting regulations may allow the agency to waive information requirements. Applicants that expect such waivers, however, do so at their own risk if they do not comply with the procedures that govern requests for such waivers and if they fail to provide themselves adequate time to collect the information should a request for a waiver be denied.

Although it may seem obvious that instructions regarding the completion of applications should be followed, permit applicants and their consultants frequently fail to read them carefully. The result is that applications are submitted on incorrect forms, lack required information, or unintentionally violate disclosure requirements. In some instances, failing to follow the instructions can unintentionally result in major gaps in permit coverage. For instance the instructions associated with the EPA's NPDES permit applications explicitly require applicants to indicate whether specific pollutants are likely to be present in the effluent regardless of whether they are expected to be removed by treatment processes. An applicant who indicates that a pollutant will not be present based on the assumption that it will be removed by a treatment system will not only violate the instructions, it will also inadvertently conceal information that the agency would otherwise use to consider whether effluent limits are needed. Although the applicant may have no intention to conceal the pollutant, such action could lead to a determination that the pollutant is not covered by the permit.

Environmental professionals need to ensure that they have all the instructions and guidance available before preparing an application. In particular, one should be aware that agencies sometimes express application requirements in instructions attached to application forms, in regulations, and in guidance documents. For example, the EPA has provided guidelines for preparing applications for permits issued under the NPDES program in the NPDES regulations, in the instructions accompanying the forms, and in the instructions contained in the forms.[75,76] Although these instructions are roughly consistent, no single set describes all the information necessary to complete an application. Furthermore, important guidance to be considered when completing applications for complex sources can be found in such documents as the *Technical Support Document for Water Quality-Based Toxics Control*.[77]

In many instances, problems with permits arise because the applicant unnecessarily forces the permit writer to rely on incorrect or insufficient information. One way to minimize such problems is to ascertain any issues that may require the submission of information beyond that specifically required by the application. Agencies are usually willing to meet with applicants to discuss whether additional information may expedite the permitting process. For example, if the agency anticipates that the permit will contain effluent limitations necessary to ensure compliance with water quality standards, the permit applicant may want to collect additional monitoring data to allow the agency to consider issues associated with effluent variability that may result in lower limits.

Another consideration when completing applications is how the information will be comprehended and used by the permit writer. Writing permits can be a very human process, and environmental professionals need to make allowances when preparing applications. A common problem is that persons who draft applications presume that the person who is assigned to write the permit will be knowledgeable about the facility and its activities. Permit writers must work with the information provided by the applicant and cannot be expected to discern information that is not adequately expressed. When presented with inadequate information, permit writers will typically respond by requesting that the applicant supply additional information or will prepare the permit using the information they have. In the first instance, providing insufficient information to the permit writer can slow the permitting process. In the latter instance, a shortfall of information may result in inappropriate permit conditions.

The level of detail is an important issue that deserves thoughtful consideration when preparing an application. The problem with being too general is that it can cause the agency to overlook issues that must be addressed by the permit. In some cases, unnecessary detail can result in permitting problems, such as unreasonable constraints or unduly limiting the coverage of the permit. As a general rule, permit applicants should not be overly specific or too general when preparing applications but instead should try to provide the level of detail indicated by the instructions. If it is difficult to ascertain what level of detail is required to meet the application requirements, the applicant should ask the agency to provide written guidance describing the detail required.

One problem with unnecessary detail is that many permit writers have the tendency to incorporate text verbatim from applications when writing permits which can lead to unreasonably narrow requirements when applications contain information that is more detailed than needed to convey relevant information. For example, an applicant who specifically describes in the application that the rinse tank is a "Model 200R manufactured by Johnson Tank Corporation" may find that the resulting permit is written in a way that prohibits the permittee from using other brands and models (e.g., "parts shall be rinsed in a Johnson Tank Company Model 200R rinse tank"). These situations present compliance problems when the applicant wants to change equipment, because it is difficult to assess when the agency genuinely intends to make the use of specific equipment a compliance requirement. Overly specific descriptions can also result in interpretations that exclude certain equipment and activities from coverage under the permit. For instance, describing emissions and discharges solely in terms of routine operations may result in an agency determination that intermittent emissions and discharges, such as those occasioned by maintenance activities or shutdowns, are not covered by a permit.

Another factor that environmental professionals should keep in mind when preparing applications is that the agency staff who will process the application may understand little about the activities for which coverage is sought. Clarity is important, and the application should not be compromised by using industry jargon to describe equipment and processes when such terms are unlikely to be understood by persons outside the industry. One way to assess how much the permit writer knows about the activities to be permitted is to meet informally to discuss the permitting concerns. Note, however, that applicants must remember that oral briefings do not negate the need to follow the application procedures, as such briefings will not be part of the administrative record. In any case, the application should be prepared in a way that clearly describes the operation to be permitted to anyone who reviews it because the ultimate audience may be broader than the permit writer. For example, if the permit is appealed, the application will constitute an exhibit considered by the administrative law judge.

Another important consideration in preparing permit applications is to use sound technical analyses when developing estimates of parameters such as emission or production rates. For example, parties should avoid excess conservatism when estimating air emissions, as overestimating in one context may constitute underestimating in another. Because it is difficult to predict all the regulatory matters that may be affected by estimates, the better long-term policy is to use the best technical basis for developing estimates and avoid attempts to bias any one estimate to achieve a particular objective. Otherwise, a pattern of inconsistent estimating practices may develop which will damage a party's credibility with agencies. When in doubt as to how to present estimates, it is sometimes better to provide a range of estimates along with a description of how they were developed.

8.4 COMMENTING ON DRAFT PERMITS

Permit applications are important because they provide the information that the agency uses to prepare draft permits, but it is not until the applicant sees the draft permit that it will know exactly how the agency intends to regulate the activities covered by the permit. Once the draft permit is issued, applicants can choose not to provide comments, in which case the permit will be issued in final form without modification unless some other party raises issues during the comment period. The applicant can also choose to submit comments which the agency is required to review and consider. Should the agency believe that the issues raised in the comments warrant modifying the draft permit, such changes will be reflected in the final permit. Even if the agency decides it will not modify the permit, it must explain its reasons for not doing so. Because the opportunity to comment on draft permits gives applicants their best opportunity to influence the requirements set forth in the permit and how they are expressed, serious attention should be given to the review of all draft permits.

Once the agency issues the draft permit, the applicant and the public will be given a limited time in which to comment on its provisions. Failing to raise issues during the comment period will usually preclude parties from raising them in future challenges to the permit provisions.[78] Because, as a practical matter, it is difficult to persuade an agency to modify a permit after it has been issued in final form, the submission of comments often provides interested parties their best opportunity to influence the content of pending permits. If more time is needed to prepare comments, an applicant can request the agency to extend the comment period. When reasonable in scope and made sufficiently early during the comment period, agencies usually grant extensions. If there is sufficient public interest in a permit, the permitting agency may conduct a public hearing at which comments are presented orally; however, public hearings generally provide limited opportunity to present substantive or detailed material, and most efforts should be directed toward preparing written comments.

Even when a comment period is extended, the time available to prepare and submit comments on a draft permit can be too short to line up the expertise and resources necessary to review a permit properly; therefore, the applicant should prepare for the comment period before the agency issues the draft permit. Major permits, particularly those that contain effluent limits based on water quality standards, can be very complicated and may entail issues encompassed by several disciplines. To ensure that they are able to review a pending draft permit comprehensively, applicants should arrange to have the necessary talent available when the draft is issued. Assembling a formal review team may be appropriate for major permits, such as those issued to large industrial facilities. The review team should be made up of persons who are expert in the major areas regulated by the permit. For example, major air and wastewater permits often entail requirements that require expertise in the areas of engineering, analytical chemistry, and law if they are to be reviewed adequately. It can also be useful to have the reviewers meet to discuss their issues and concerns, as the discourse between persons with different areas of expertise leads them to discern issues that its individual members otherwise would not have noted.

Having established who will be responsible for reviewing a draft permit, organizations need to address how they will conduct the review and prepare comments. Because the purpose of commenting is to influence the legal effect that the permit will have on the organization, it is important that all the relevant issues be communicated in writing to the agency so that they are entered into the administrative record. Although it can be useful to discuss issues associated with the draft permit with agency staff, oral communications do not appear in the administrative record nor do they bind agencies as a practical matter. Therefore, it is important not to rely on assertions that the agency will use its discretion and not enforce unreasonable provisions. Furthermore, even if an agency refrains from enforcement, unreasonable provisions can still be enforced by environmental groups in citizen suits, so even if an issue appears to resolved orally the applicant must still document the issue and the contemplated resolution if it wants to ensure the issue is actually addressed in the final permit.

Organizations should also take care to review draft permits and determine the important issues early in the comment period. When reviewing permits, organizations should not only consider whether the provisions conform with legal requirements, but they should also evaluate the feasibility of compliance. It is also important to scrutinize the entire permit, rather than focus on the provisions that tend to dominate attention, such as emission and effluent limitations. Often, provisions relating to items such as monitoring and reporting can lead to significant compliance problems if not resolved before the final permit is issued. For example, a requirement to submit monitoring information by the 15th day of the next month will not allow for compliance if the necessary analytical results are not available until the 20th day. Issues such as these are often easy to resolve pending issuance of the final permit but difficult afterwards as agencies often do not want to devote resources to these kinds of issues. However, while it may seem unlikely that late monitoring reports would themselves constitute grounds for enforcement, such violations would probably be incorporated into an action motivated by other reasons and would be used to portray the permittee as indifferent to compliance.

The provision in the permit should be evaluated to ensure that the permitting agency has the legal authority to incorporate them into the permit. In general, the agency will have some authority to add non-standard provisions to a permit but must be able to establish why they are needed to effect a legitimate regulatory purpose. Permit writers, though, do not have the authority to amend regulations and thus cannot negate provisions required by regulation such as signatures, modification requirements, and permit reopener provisions. Any provisions that is set forth in the permitting regulations must be incorporated into the permit as a matter of law.

The agency's regulations and guidance prescribe the procedures and issues that permit writers are required to follow or consider but it is not unusual for permit writers to disregard them when preparing a permit. With regard to some issues, the permit writer will have discretion to exercise his or her best professional judgment. In other cases, this discretion will be significantly constrained by formal policies or practices. For example, the NPDES permitting rules provide that limitations on internal wastestreams may be used only in exceptional circumstances, which must be explained in the fact sheet accompanying the draft permit. Thus, if the permit writer proposes internal limits or restrictions on the discharge of internal wastestreams, the applicant may want to question the basis for doing so.

Emission and effluent limitations that are based on modeling assessments deserve special scrutiny. Agencies are increasingly relying on modeling to assess the effects that emissions and dischargers have on environmental quality. In the typical case where modeling is used, the agency will evaluate the predicted concentrations of pollutants against environmental quality criteria or assessed risk values. Both the modeling and selected criteria involve the exercise of judgment and the potential for inappropriate or erroneous decisions regarding how emissions and discharges should be evaluated. Although environmental modeling is often shrouded with an aura of technical sophistication and mystery, many models are based on relatively simple equations and are not particularly effective at predicting the true transport or fate of pollutants. When modeling results in troublesome conditions, applicants may want to evaluate the assumptions underlying the modeling and submit comments. Agencies are sometimes open to modifying the initial assumptions underlying the modeling and may incorporate revised modeling results into final permits. Even when an agency declines to address modeling issues, the submission of comments preserves the modeling issues for possible resolution in a permit appeal.

Measurement is one of the most difficult issues associated with environmental compliance, and applicants are well advised to pay special attention to monitoring and analytical requirements in proposed permits. Draft permits usually specify the analytical methods that may be used to fulfill the requirements of the permit, but if they do not then the applicant may want to address the issue in the comments. In most cases, the draft permit will incorporate the analytical methods by reference but sometimes are identified directly in the permit. Regardless of how the analytical methods are defined, they are extremely important because the permittee will not be able to lawfully use an alternate method unless the permit is modified. Furthermore, courts have ruled in enforcement cases that permittees cannot contest the validity of data results submitted in their monitoring reports which means that the forced use of an inappropriate method could create a substantial compliance problem.

Although monitoring is a sensitive regulatory issue, permit applicants should not assume that the analytical methods specified in the permit are suitable for the matrices that will be tested. Draft permits often specify methods that are not suited to the intended application or that specify irrational resolution and quality control requirements. For example, a draft permit might specify that analyses meet a detection limit that is unreasonable for particular kinds of industrial wastewaters that will be tested. Also, if a draft permit fails to specify an appropriate method altogether, serious disputes can arise over issues such as detection limits, cost, and reliability. When a permit incorporates analytical requirements by reference, the underlying methods should be assessed for their applicability and suitability. Many of the methods approved by agencies for general use cannot be used to analyze certain kinds of industrial emissions and wastewaters because they are highly subject to interferences. If a permit requires monitoring for such parameters, the applicant should ascertain whether the proposed analytical methods can provide accurate measurements. If not, the comments should propose modified or alternate analytical methods. Similarly, if an unconventional analytical method is proposed in a draft permit, it should be reviewed by a person qualified to assess whether the method is appropriate for the intended application.

Unless the issues of cost and availability are considered before the final permit is issued, applicants may find themselves faced with difficult and expensive compliance issues. Applicants should estimate the projected cost of meeting the analytical requirements in the draft permit to determine whether they are reasonable. In some instances, permit writers may impose monitoring requirements without knowing what they will cost. By estimating such costs, permit applicants can raise the issue before the final permit is issued and propose either a different analytical method or a reduced monitoring frequency. The permit applicant should also investigate the availability of laboratories that can perform the test method. For example, some bioassay protocols can only be performed during certain months of the year. Similarly, the competitive nature of commercial laboratories has caused many to abandon obscure methods that require specialized or custom-fabricated equipment. Although it may be possible to contract with a university or research institute for such analyses, the cost and reliability of service may not be at desired levels. Problems with availability can be supported in comments by providing the addresses and telephone numbers of laboratories who can attest to such problems as insufficient availability and lengthy turnaround times.

An illustration of the degree to which the analytical and enforcement communities can ignore widespread noncompliance with regulatory requirements is presented by the quality control provisions of the test for 5-day biochemical oxygen demand (BOD). The way in which the BOD methods were officially adopted also illustrates the sometimes haphazard approach that the EPA has taken to prescribing the analytical procedures that regulated parties must follow with regard to permits.

Case Study: 40 C.F.R. § 136.3 and Biochemical Oxygen Demand

Environmental professionals often assume that the likelihood of finding substantive issues associated with traditional analytical methods referenced in permits is remote; however, most referenced methods have been developed outside of regulatory contexts and are often expressed in ways that serve regulated parties poorly when used as regulatory conditions. For example, biochemical oxygen demand (BOD) has been a parameter used in the water quality field for over 100 years. Its importance as a regulatory parameter is evidenced by EPA regulations that require most wastewater discharge permit applications to include monitoring data using the BOD test methods approved by the agency at 40 C.F.R. § 136.3. Although various test methods for BOD have been in effect for decades, few people are aware of their inconsistent and irrational quality control requirements.

The EPA has approved five methods for determining BOD, and most wastewater discharge permits issued in the U.S. require their use. The two most common are method 5210B as described in the 18th edition of *Standard Methods for the Examination of Water and Wastewater* and the 1983 revision to EPA method 405.1 which adopts by reference the procedures set forth in method 507 of the 15th edition of *Standard Methods for the Examination of Water and Wastewater.*[79,80] Two less common methods are method 973.44 of the Association of Official Analytical Chemists that follows procedures close to those set forth in the 14th edition of *Standard Methods for the Examination of Water and Wastewater* and method I-1578-78 of the U.S. Geological Service (USGS) which essentially prescribes the use of any general method for determining BOD but endorses the one set forth in the 14th edition of *Standard Methods for the Examination of Water and Wastewater.*[81,82] The fifth method is described in a document titled *American National Standard of Photographic Processing Effluent* published by (but no longer available from) the American National Standards Institute (ANSI).[83]

The history behind EPA approval of these methods is convoluted but reflects a process of adoption by reference without seriously considering the quality control requirements. The first approved BOD method (1973) contained quality control guidelines but no rigid criteria. The agency replaced this method in 1976 with the BOD method set forth in the 14th edition of *Standard Methods for the Examination of Water and Wastewater* and added the USGS and ANSI methods as other options. These methods likewise refer to quality control guidelines but impose no rigid criteria. In 1984, the EPA amended the regulations to include the 1979 version of EPA method 405.1 as an approved method but surprisingly did not approve the 1983 version. At the same time, the EPA amended its regulations to include the recently updated method 507 described in the 15th edition of *Standard Methods for the Examination of Water and Wastewater* which requires tests to be rejected if the glucose-glutamic acid checks do not fall within a specified range. This quality control requirement was made more strict when EPA amended its regulations in 1991 to replace method 507 with the updated *Standard Methods for the Examination of Water and Wastewater* method 5210B. However, at the same time, the EPA amended its regulations to incorporate the 1983 version of method 415.1 which adopts method 507 from the 15th edition of *Standard Methods for the Examination of Water and Wastewater* by reference.

As things stand, three of the approved BOD methods require laboratories to consider quality control guidance but do not require tests to be rejected, while two of the methods contain different but rigid quality control criteria that require nonconforming test results to be rejected. Furthermore, the inconsistency of these regulatory-requirements-by-reference is compounded by the irrationality of the quality control criteria of method 5210B and the 1983 version of method 405.1. Although the standard practice is to set quality control criteria for laboratory testing based on a range determined by the mean ± 2 standard deviations of the quality control checks, the ranges of acceptable control limits specified in *Standard Methods for the Examination of Water and Wastewater* method 5210B and EPA method 405.1 are based on the mean ± 1 standard deviation of data collected in intra-laboratory performance evaluations conducted by the EPA. Neither *Standard Methods for the Examination of Water and Wastewater* nor the EPA has stated a rationale for setting the control limits for individual laboratories based on the mean ± 1 standard deviation. However, basic statistical theory would indicate that approximately 32% of the data determined by these methods will fall outside their respective control limits and therefore should be rejected. Prior to 1984, the EPA had approved no BOD methods that required laboratories to reject tests based on glucose-glutamic acid checks but instead required them to compare the results of the glucose-glutamic acid checks to a table of expected values and to check the seed source and technique if the results differed appreciably from guideline values. While it appears probable that the EPA did not intend to make the quality control criteria in method 5210B and the 1983 version of EPA method 405.1 into legal requirements, it certainly cannot be said that it gave the issue the attention it deserves. The lesson to be learned is that organizations should not assume that even longstanding regulatory requirements referenced in draft permits will not present compliance issues.

Because the failure to comply with reporting requirements can be very costly to permittees, applicants should be careful to ensure that the proposed reporting requirements do not impose unreasonable or unclear burdens. For example, some permits require that upsets or malfunctions be reported but do not clearly define what is meant by those terms. Similarly, the permit should address how reports should be made when required during weekends, holidays, and other periods when the government offices are not staffed. Many agencies now use voice mail or call forwarding during such periods to increase their access during off hours. Some organizations, though, may prefer to make notifications using facsimile, which does not make the information as accessible. To avoid the potential for disputes over the adequacy of communication, these kinds of issues should be addressed in the comments.

The reporting of monitoring data is another issue that deserves detailed attention. For instance, many permits require the applicant to report averages of monitoring data but provide no guidelines for how analytical data should be averaged when the data include results that are below detectable concentrations. Although there are several means to address these situations, it is better to reach agreement with the agency before the final permit is issued than to wait until a critical compliance determination hinges on the method used to average data. Furthermore, some parameters are not readily averaged, such as pH, which is reported on a logarithmic scale, and fecal coliform when measured using the most-probable-number (MPN) method. If the draft permit does not specify whether and how these results should be averaged, the issue needs to be discussed in the comments. Applicants should also be careful that the draft permit does not propose open-ended requirements such as "appropriate statistical techniques" to assess compliance with emission or effluent limits. Unless such requirements are stated with certainty, they are certain to cause compliance problems.

Because compliance with permit conditions will often depend on the efforts of facility managers and other employees who may not be environmental specialists, organizations should use comments to address provisions that are difficult to comprehend. Furthermore, poorly expressed terms in permits have the potential to create significant compliance problems if an agency later interprets the requirements in a heavy-handed or unusual manner. Some commonly used terms that lend themselves to ambiguity include "appropriate," "timely," "*de minimis*," and "significant." When such words are used in a context necessarily prone to subjective interpretation, the permittee should request that the permit be revised to express its requirements more clearly. Agencies may also use the definitions section of the permit to put an unusual gloss on permit terms. Because distorted definitions can lead to misunderstandings about permit requirements, the definitions section of the permit should be carefully reviewed to ensure that its terms do not impose unintended or irrational obligations.

Of particular concern are provisions in permits that are inadequately defined. For example, a permit that expresses an emission limit in terms of tons per year expressed as an annual rolling average can be interpreted in several ways. In theory, the term could be construed to mean a 365-day average, 52-week average, or 12-month rolling average, although none of these averages would normally be expressed in terms of tons per year. It is possible that an agency that would draft such a provision intends the average to encompass an interval of several years. Although the issues raised by poorly defined terms tend to be subtle, addressing them during the comment period will sometimes be the only means to avoid noncompliance problems later.

Once the applicant has identified the issues that need to be addressed in comments, it needs to determine how to express them. Comments tend to be most effective when presented in a consistent format, as agencies are better able to discern the issues. For example, the applicant can organize the issues into chapters which are further divided into a background section, a short statement of the issue, a discussion of the issue, and proposed text for modifying the permit. Regardless of format, the applicant should cite the scientific, legal, or other material supporting its position and provide these materials if the agency does not have ready access to them.

8.5 APPEALING PERMITS: SOME PRAGMATIC CONSIDERATIONS

Although it is rare for an agency to modify permits in ways that address all the permittee's concerns, in most cases the issues are sufficiently resolved and no further action is taken to amend the terms of the permit. However, the regulations that govern permit appeals usually require the permittee to file a notice that it intends to pursue an appeal within a limited period which is rarely longer than 30 days and in some cases considerably shorter, thus it is critical for permittees to review the permit immediately after it is received in its final form. The sensitive issues associated with a permit will usually have been identified during the review of the draft permit, but a full review is warranted because agencies occasionally modify permits after public comment in ways that may not have been anticipated by the permittee.

After the permit has been reviewed, the permittee must decide whether to go forward with an appeal and formulate a strategy for prosecuting the appeal. Remember that if the permit is left unchallenged, its provisions become fully enforceable on the effective date of the permit. If the permit has been issued pursuant to a statute that provides for citizen suits, such as the Clean Air Act or Clean Water Act, private parties will be able to sue to enforce the permit and to recover penalties. Because a bad permit provision can expose the permit to considerable liability, an appeal may be necessary to avoid enforcement at a later date. The appeals process can be used to challenge provisions that have no rational basis or that the agency had no authority to incorporate into the permit. The permittee can also challenge any actions by the agency that did not comply with the procedural requirements set forth in the permitting regulations. In most cases, a permittee has little or no chance of prevailing on issues that the agency is required by law to incorporate in the permit.

Some permittees are reluctant to appeal permit provisions because they fear that they will alienate government agencies and incur large legal expenses. Although these risks exist, they usually do not evidence themselves in the typical appeal because most issues can be resolved without a hearing. Agencies are unlikely to be offended by the filing of an appeal if the underlying motive is to avoid being forced into noncompliance rather than delaying compliance expenditures. So long as the permittee has a legitimate reason for pursuing an appeal, they should not fear retaliation by the government. Furthermore, such retaliation is illegal under the U.S. Constitution and can subject the agency and its employees to damages.

Administrative tribunals and courts tend to defer to agency legal and technical positions on matters within their area of responsibility. Unless the matter involves issues where the agency clearly has no rational basis for its position, it is generally prudent to try to resolve the appeal through negotiation. In fact, one benefit of appealing a permit is that the administrative burden of defending its decisions tends to make agencies more receptive to reasonable resolutions of permitting issues. To maximize the opportunities to resolve an appeal without a hearing, the parties should be careful to avoid uncivil conduct or tactics. Instead, they should explore the opportunities to negotiate on the contested issues and to find resolutions that meet their respective needs. Not only does this approach tend to increase mutual respect, but it also helps to reduce legal expenses.

An important issue to consider is the effect that the appeal will have on the contested provisions. The regulatory schemes for issuing permits vary in regard to this issue, but in some cases an appeal will trigger an automatic stay of the contested provisions. In other cases, the contested provisions may go into effect unless the agency, tribunal, or court issues a stay. Because obtaining a stay may require specialized knowledge and take some time, it is best to consult with legal counsel prior to the issuance of the final permit.

For most permits, the appeal should focus on technical issues even though the process is a legal proceeding in which it is almost essential to retain counsel to prosecute the appeal. Many agencies allow parties to represent themselves or allow advocacy by parties other than attorneys; however, such representation is generally a mistake because the failure to follow procedures or meet the standard of proof can result in the permittee losing for reasons other than the substantive merits of the appeal.

Once the decision is made to appeal a permit, it is essential that the appeal procedures be followed meticulously, because appellate tribunals have the disconcerting tendency to dismiss cases based on the slightest misstep. If the appropriate notices are not filed within the period allowed, the tribunal is unlikely to consider the matter. In addition, appeals are typically decided on the administrative record, which should be obtained as soon as possible following the decision to file an appeal. The best way to obtain these records is to request a copy of the administrative record pursuant to the Freedom of Information Act or the analogous state government records statute.

8.6 IMPLEMENTING THE FINAL PERMIT

Once the permit is issued, it is important to implement the measures and procedures that will be needed to ensure compliance. The basic objective is to communicate the key requirements of the permit to the individuals who are responsible for the affected activities. The first step, which should already have been completed during the application and comment process, is to assess what activities are affected by the permit. Once this has been done, the individuals whose responsibilities will be affected by the permit can be identified. When these individuals are identified, the organization can optimize its communication of the permit provisions.

Although key personnel such as plant managers and environmental staff should read and become intimately familiar with all the permit conditions, many employees will have their duties affected in limited ways. To improve comprehension and avoid information overload, it is sometimes better to provide these employees with just the information that is relevant to their duties. For example, a maintenance director may need to know that the permit requires that the pollution control equipment be serviced according to manufacturer recommendations but probably will not need to know about the requirements for quality assurance in the laboratory. When in doubt, though, organizations should err on the side of providing more information.

When necessary, to enable comprehension, the organization should rephrase and expand on permit conditions when providing information to employees. Permit conditions are sometimes written at a level that will be difficult for employees to comprehend. Conveying such information in simplified text can significantly improve the level of understanding throughout the organization. Similarly, some permits incorporate regulations by reference, which means that employees who do not have access to those regulations may miss important requirements. In such cases, the organization should either provide a copy of the regulations or summarize the requirements expressed therein.

Another important task in implementing the permit is to assign responsibility for compliance and reporting obligations to individual employees. Unless individuals are told that they are responsible for ensuring compliance with certain provisions, organizations may find that inadvertent noncompliance occurs because responsibilities were misunderstood. Certain activities, such as reporting and inspections, tend to be overlooked unless they are assigned to specific individuals. For example, it is not unusual for plant employees and corporate environmental staff to concurrently assume that the other group has assumed responsibility for activities such as reporting.

8.7 PERMIT MODIFICATIONS

It cannot be overemphasized that agencies generally disfavor permit modifications because they impose a significant administrative burden on typically strained resources.[84] Processing a permit modification can require almost as much effort as reissuing a new permit. Furthermore, because a decision whether to modify a permit is within the complete discretion of the agency, permittees have practically no leverage to force the process. Nonetheless, environmental matters are sufficiently complicated that permittees sometimes need to have permits modified to address unforeseen developments or issues that were not considered during the permitting process.

There are several reasons why permittees should not rely on the modification process as a primary means of addressing permitting issues. First, obtaining a modification is a lengthy process and rarely provides a quick solution to a compliance problem. Second, a request for a modification often publicly highlights a serious compliance issue that may subject the permittee to enforcement by any number of parties. As a practical matter, the prospect that an agency may deny the modification request means that the permittee may find itself more vulnerable to enforcement for having tried to resolve the issue. Third, permit modifications require effort on the part of agencies which may insist on concessions in exchange for the modification. Although situations do arise where it makes sense to request that a permit be modified, prudent permittees should make all possible efforts during the permitting process to avoid the need for modifications after the permit is issued.

Agencies frequently have different procedures for processing ministerial and substantive modifications. Ministerial modifications are used to address matters that do not require reconsideration of substantive provisions. Examples of such modifications include changing the name of the owner following the sale of a facility and correcting typographical errors. Such matters can usually be addressed through simple amendments and do not require public comment. Permit modifications that affect the substantive obligations of regulated parties require the agency to expend more effort. For example, more staff time is required to obtain internal approvals and the proposed modification must be made available for public comment.

To reduce the effort associated with permit modifications, regulated parties sometimes seek what is effectively an informal modification by asking the agency to interpret away the troublesome provisions or provide assurances that no enforcement will be taken. Unfortunately, interpretations and assurances do not always protect permittees from enforcement — particularly when made verbally by lower level staff or when contrary to the plain meaning of the permit. In addition, any representation made by an agency regarding its intentions not to enforce a provision will not bind other enforcement agencies or private litigants.

Another common but more formal means of addressing problematic permit provisions is the compliance order. These orders allow regulated parties to proceed with activities that violate a permit condition if they agree to certain restrictions. However, such orders constitute neither permits nor permit modifications and do not bind parties other than the permittee and the agency that issued the order. This means that other government agencies and private citizens are free to sue for violations of permit provisions or regulations addressed by the compliance order.[32]

When a permittee determines that a permit modification is necessary, it should raise the matter informally with agency staff before making the formal request. In particular, the agency should be told why the modification is needed and the ramifications to the permittee if the permit is not modified. A good explanation for why the modification is needed will help make the agency more receptive to the subsequent formal request. Also, agencies are more likely to agree to a modification if the permittee can justify why the issues were not addressed during the permitting process. Good reasons for seeking a modification include circumstances that were not foreseeable, such as emissions or control efficiencies for new processes that could not have been estimated with certainty. Another good reason would be regulatory developments that could not have reasonably been anticipated when the permit was issued. Agencies tend to be less receptive to modification requests rooted in matters that were within the permittee's control at the time the permit was issued. For example, agencies tend to be disinclined to address issues caused by inadequate descriptions in the application or failure to comment on provisions that are not feasible or were ambiguously expressed in the draft permit.

When making a request, it is important to follow the procedures that govern permit modifications. Furthermore, anything that can be done to streamline the process will be to the benefit of the permittee. It is often helpful to propose draft text for the requested modification and to include all the readily available supporting material when submitting the request. By reducing the demands on the agency to draft text, research issues, and obtain supplemental information, the permittee can do much to expedite the modification. However, many modification procedures require public notice and comment and thus extend the time required to obtain the modification. In some cases, modifications will take several months to obtain even if the agency has no substantial objections.

9 Measurements, Monitoring, and Their Implications for Compliance

Measurements and monitoring are serious compliance issues because all the major environmental regulatory programs rely on measuring environmental parameters for their effectiveness. Without such measurements, it becomes impossible for agencies to regulate most conduct affecting the environment in a meaningful way. Measurements are used to assess compliance with emission and discharge limits, and they provide critical information regarding environmental quality that is used to develop and implement regulatory strategies. Failing to provide quality measurements with regard to parameters that are not subject to limits can nonetheless infringe on serious agency interests and expose regulated parties to enforcement.

Another factor that makes measurement a sensitive compliance issue is that measuring environmental parameters is a complicated and sometimes inexact science, as in many instances the phenomena to be measured are exceedingly large or diminutively small. Unfortunately, regulatory requirements rarely make substantial allowances for the unavoidable uncertainties associated with environmental measurements. For instance, regulations and permits typically impose limitations and monitoring requirements on parties based on the mostly unrealistic assumption that most releases can be measured with precision; however, many parameters can be assessed only by measuring parts of the whole and extrapolating those results into quantitative estimates. The result is that most environmental measurements involve the exercise of considerable judgment. Because judgment is prone to subjective evaluation, measurement issues can generate regulatory controversy. Furthermore, agencies have held regulated parties liable for using measurement information developed by the government. For example, the U.S. Environmental Protection Agency (EPA) has held companies liable for failing to obtain construction permits after emission estimates prepared using sources such as AP-42 were found not to reflect actual emissions.[85] The agency justifies enforcement under these circumstances because it believes that regulated parties should bear the risk of not being able to estimate accurately.[62]

Another reason why the government is sensitive to issues involving environmental measurements and monitoring is that most environmental programs depend upon regulated parties to collect and report environmental data to government agencies. Unless the government is comfortable that self-reported measurements are reliable, many of its regulatory programs would be rendered almost ineffective. Because monitoring and reporting practices can be abused, the government has a strong incentive to punish transgressors severely, and the majority of cases involving false monitoring or laboratory fraud are enforced as criminal matters.

9.1 SAMPLING ISSUES

Good sampling practices are essential to ensure the integrity of environmental measurements. Although sampling procedures may appear simple, the failure to address regulatory requirements can lead to unanticipated compliance problems. As a practical matter, most environmental measurements require parties to collect and analyze small fractions of the medium to be evaluated and then extrapolate to assess the quantity or concentration. Thus, the primary compliance consideration when sampling environmental media is to ensure that collected samples are representative of the whole. Unfortunately, representative sampling is not always easy, as many media associated with regulated activities are heterogeneous. For example, constituents in the sampled material can vary with time, depth, area, flow, pH, and temperature, and representativeness can be impaired by phenomena such as turbulence, quiescence, and chemical interactions. Although overcoming these obstacles is beyond the scope of this book, it is essential that issues concerning the ability to collect representative samples be addressed when compliance issues are involved.

Another issue involving representative sampling is how to address events that may contribute high levels of pollutants but are not likely to be detected by routine sampling. This situation occurs frequently in the context of air emissions because sources are mostly tested at infrequent intervals. In addition, many wastewater and stormwater sources are monitored intermittently. As a practical matter, the ability to ensure representative monitoring deteriorates as the number of samples decreases. In some instances, regulatory agencies acknowledge this issue by specifying the conditions under which sampling or testing should occur even though they may not reflect the characteristics of the releases at other times. For example, many stormwater permit conditions specify that sampling is to be done during storm events that exceed 0.1 inches of precipitation and following a period of at least 72 hours without measurable rainfall.

Many environmental permits, however, are either silent on the issue of representative monitoring or provide no guidance as to what constitutes sufficiently representative sampling. In such cases, the general assumption is that monitoring must be reasonable under the circumstances. The dilemma presented by this assumption is that reasonableness is subjective and differences of opinion can lead to disputes over whether a facility has complied with the applicable legal requirements. To some extent, a determination of an agency's formal position on what constitutes representative sampling can be gleaned from guidance documents or widely accepted reference materials. Examples of such materials include the sections on sampling in *Standard Methods for the Examination of Water and Wastewater* and the EPA's *Test Methods for Evaluating Solid Waste (SW-846).*[80,86]

Finally, it is also important that environmental sampling be executed properly. In instances such as the stack testing of air emission sources, sampling is mostly conducted by contractors who have the specialized skills and equipment to perform the required methods; however, much environmental sampling is performed by technicians who have limited backgrounds in the environmental sciences. It is absolutely critical that the persons who conduct sampling be well trained concerning the media they will be sampling. Overlooking seemingly minor matters can result in significant monitoring problems. For example, unless technicians are instructed on whether or not to collect grab samples of water from the surface, the presence of surface scum or oily sheens can substantially affect the monitoring results. Furthermore, inconsistent monitoring procedures by technicians can lead to perplexing monitoring data. Any organization that relies on its employees to collect samples for environmental monitoring should adopt formalized training requirements and prepare detailed procedures to guide sampling activities.

9.2 ANALYTICAL ISSUES

Analysis of chemical and physical parameters presents the issues of whether the analytical procedures comply with regulatory requirements and whether they produce accurate measurements. Unfortunately, this is also an area where many environmental professionals fail to appreciate the regulatory and technical issues to the extent needed to ensure total environmental compliance. To become competent in dealing with analytical issues, environmental professionals need to know more about analysis than the numeric designations, acronyms, and nicknames that are used to identify test methods. All environmental professionals need to understand that regulatory requirements determine which methods can lawfully be used for specific applications and that technical suitability by itself is not sufficient to ensure compliance. Furthermore, analytical chemistry is a scientific discipline and should be appreciated as such. In other words, better environmental professionals treat the providers of analytical services as a resource and not as vendors of a commodity to be procured at the lowest price.

9.2.1 Test Methods

To ensure that environmental data used for regulatory purposes are measured consistently and reliably, the EPA has identified specific requirements for the analytical methods to be used to determine the physical, chemical, and biological characteristics of wastewater, drinking water, air emissions, and solid waste. These methods are generally referred to as approved, official, or reference methods, depending on the regulatory context. Although many methods are widely accepted by the scientific community, the EPA usually mandates that only approved methods be used to fulfill regulatory requirements such as wastewater monitoring or air emission testing. In addition, EPA policy has been that methods approved for use in one program cannot be substituted for the methods approved for another program, irrespective of whether those methods are suitable for the matrix to be analyzed.[87]

While the EPA has explicitly expressed the requirement to use approved methods, it has sometimes been less than diligent in its efforts to provide the analytical requirements. For example, the EPA issued its *Methods for the Chemical Analysis of Water and Wastes* in 1979 and approved its methods for use in complying with the requirements of the National Pollutant Discharge Elimination (NPDES) program.[88] The agency issued a revised version in 1983 but did not officially approve the version for NPDES monitoring until 1991.[89] Although there does not appear to have been any intent on the part of the agency to withhold approval of the 1983 version, its actions regarding analytical issues are inconsistent with its expressed intent that regulated parties use approved methods for environmental monitoring.

Unfortunately, the EPA's sometimes capricious attitude towards defining analytical requirements has misled some parties into believing that they have more latitude in selecting test methods than is technically allowed by the law. In most instances, these parties do not intend to violate legal requirements but rather misunderstand that agencies often consider these issues important. Typically, most analytical chemists select analytical methods from a scientific rather than a regulatory perspective. Because there are many unapproved methods that perform as well as the methods that the EPA has formally approved for its programs, there is little scientific justification not to use them; nonetheless, the EPA has voiced concern that the use of such methods can violate environmental laws.

Despite the seemingly clear prescription to use approved methods, there are sound technical reasons why regulated parties sometimes fail to do so. For instance, parties may use unapproved methods or unauthorized modifications to approved methods to resolve interference problems or to increase the sensitivity of an analysis. Other reasons may be to reduce costs or to expedite analytical procedures. Although such modifications may degrade method performance, in many instances the extra performance is not needed to measure analytes at the levels of regulatory concern.

Another explanation for why regulated parties use unapproved methods is that government agencies have been complacent about enforcing such requirements. Most agencies do not carefully review submissions from regulated parties to verify that approved methods were used to develop the information being reported. Even when agencies observe that approved methods have not been used, they typically disregard the issue and allow regulated parties to continue using unapproved methods. Nonetheless, there are good reasons to ensure that approved methods are used when legally required. First, the failure to follow regulatory requirements associated with test methods can cause parties to incur large numbers of violations, as many programs require frequent testing. Second, there are few defenses available to parties that use unapproved procedures. Although a regulated party is free to question the wisdom of any regulation, the fact that a regulation is unreasonable or even irrational will rarely excuse noncompliance, thus disregarding even irrational or unnecessary requirements can expose an organization to liability. Even permission obtained from agency staff to use alternate methods will not constitute a defense if the formal procedures for obtaining such approval were not followed.

Needless to say, organizations should ensure that they and their contract laboratories use approved methods when legally required to do so. When procuring laboratory services, parties should specify that analyses that will be used for compliance reporting purposes must be done using an approved method; otherwise, there is a significant possibility that laboratories will use unapproved methods. Perhaps the most common problem is the use of methods approved under the Resource Conservation and Recovery Act (RCRA) for analyses required by the Clean Water Act. Although the RCRA methods are scientifically sound, the EPA has stated that such methods are not substitutes for the often very similar methods approved under the Clean Water Act.[87] Regardless, analytical laboratories will not know which methods to use if the customer does not identify the purpose underlying the analysis.

9.2.2 Avoiding Analytical Problems

Because most environmental measurements involve technical and legal considerations, problems will occur if a regulated party does not affirmatively act to prevent them. In the absence of careful attention, regulated parties can encounter situations such as missed sampling rounds, lost samples, and questionable analytical results. Many analytical problems can be attributed to carelessness or ignorance, such as allowing sample holding times to be exceeded or using incorrect methods. In addition, without an effective compliance program, sloppy or dishonest employees can engage in inappropriate conduct for extended periods without being detected. All of these problems, even if unintentional, can result in serious compliance problems.

The best way to avoid troublesome results is to minimize the occurrence of these problems. Regulated parties that conduct their own monitoring should implement the procedures necessary to ensure compliance with legal and quality control requirements. Strong quality assurance and compliance programs are essential to avoid regulatory problems over the long term. In general, a sound quality assurance program will incorporate statistical control techniques, calibration procedures, and some form of auditing. Such programs can also incorporate a quality assurance policy that sets forth the commitment to quality and the respective responsibilities within the organization. While the purpose of a quality assurance program is to ensure that quality of measurements remains high, the intent of a laboratory compliance program is to ensure legal compliance and detect faulty or fraudulent practices if they occur. Without a compliance program, honest intentions on the part of management may be insufficient to prevent fraudulent laboratory practices.

Laboratory fraud has become of great concern to environmental agencies, and it is now routine for suspected incidents to be handled as criminal matters. For example, unethical employees may surreptitiously omit procedures or fake data to reduce their workload. Furthermore, there have been instances where dishonest employees have falsely claimed that management condoned or ordered improper acts. By developing laboratory compliance programs, organizations increase their ability to prevent illegal practices or detect them in their early stages. As a general proposition, a laboratory compliance program should be established using the principles for environmental compliance programs described in Chapter 4. However, the special nature of laboratories requires that other aspects be considered. In addition to basic compliance program principles, laboratories should focus on the issues of employee qualifications, training, internal reporting, and management of laboratory notebooks. Extra attention paid to these areas can enhance the ability of the compliance program to reduce internal problems.

As is the case with any activity related to environmental compliance, employee issues are critical. Organizations need to assess what qualifications are required by employees who have analytical responsibilities. With regard to laboratory work, qualifications will vary depending on the nature of the tasks involved, the analytical work, and the level of responsibility. For example, lesser qualifications are needed for a technician who performs standardized procedures on well-defined matrices, whereas considerable expertise may be required when using complex equipment to analyze complex matrices. Training of employees who conduct analyses is also important. Not only should training cover matters such as procedures and materials storage, but it should also ensure that employees understand the organization's environmental policies and other aspects of the compliance program.

Analytical work is complex and readily subject to misunderstandings and sometimes outright abuse; therefore, organizations that perform environmental analyses should address both issues when developing compliance programs. The issue of employee misunderstandings is relatively straightforward with regard to compliance programs and is most effectively addressed by focusing on communication. First, employees should be encouraged to communicate freely any issues regarding proper laboratory procedures. Although policies are useful, as a practical matter the attitude that management exhibits towards employee queries will determine how effectively the organization addresses employee misunderstandings. Managers should train themselves to listen carefully to any question regarding laboratory procedures and avoid reactions such as impatience or ridicule. Second, the laboratory should ensure that questions get addressed to qualified persons. Laboratory errors due to employee misunderstandings have a way of multiplying when employees consult with coworkers who likewise misunderstand the procedures at issue. Typically, this happens at the technician level when one low-level analyst asks another about how to perform a test. One way to encourage analysts to obtain correct information is to require them to consult with a designated manager when questions arise.

Outright misconduct by employees presents serious compliance issues, and laboratories should take reasonable measures to prevent and detect it. Sometimes employees are tempted to gloss over problems or fabricate information, reasoning that no one will discover the misconduct. However, false reporting is difficult to conceal in the face of efforts to expose it, which means that government agencies and laboratory customers sometimes discover employee misconduct before laboratory managers do. Common means by which false reporting is discovered are whistleblower complaints, laboratory audits, and comparisons to data from other laboratories.

Laboratories should be sensitive to temptations to cover up problems and take measures to minimize the possibility that employees may act on them. For example, laboratories should establish policies and procedures that specify how and when employees are to notify managers about problems. Laboratories should also adopt procedures intended to reduce problems associated with sloppy or misunderstood testing procedures such as supervision and monitoring of laboratory procedures and the frequent review of laboratory notebooks.

Laboratory notebooks can constitute a powerful element of a compliance program. Notebooks are commonly used because they provide a convenient way to record and store analytical information as it is being developed. Because these books can be used to reconstruct a history of laboratory practices, they usually are critical evidence during government investigations of laboratories suspected of misconduct. In fact, notebooks are usually among the first items subpoenaed or seized by the government when investigating allegations of criminal activities by laboratories. In such cases, the government investigators will scrutinize the notebooks or bench sheets for signs of misconduct such as altered entries, missing data, or substituted pages. On the other hand, when properly prepared, laboratory notebooks can be used to show that an honest laboratory did not knowingly engage in improper activities and that fraudulent practices did not occur. Laboratories can thus use laboratory notebooks to prove their honesty, but to do so they must establish and implement the necessary procedures before they come under scrutiny.

Rigorous procedures for the use of laboratory notebooks can reduce the exposure of a laboratory and its managers to liability, particularly in cases where a staff analyst covertly fails to conduct analytical work properly. The key elements of maintaining laboratory notebooks for compliance purposes include: (1) ensuring that they contain sufficient information to document each important aspect of an analysis, (2) keeping them in a form that minimizes the opportunities for subsequent alterations, and (3) inspecting the notebooks at frequent intervals to ensure that the entry procedures are followed and to detect irregularities. Laboratories should implement formal procedures that describe how data are to be entered into notebooks and how the notebooks will be reviewed and stored. It is recommended that laboratory managers personally train newly hired laboratory staff in the data entry procedures specific to the laboratory and document the training in the employees' personnel records. This will help to emphasize the importance that the laboratory attaches to the notebook procedures. It is also recommended that employees immediately receive refresher training whenever circumstances suggest that more training is needed.

At a minimum, the laboratory notebook should contain the following information for each analysis: the signatures of the analyst and any assistants, the date and time of each analysis, all information used as input for calculations, and an indication of whether any problems or anomalies were associated with the analysis. The purpose of this information is to provide a clear record of who conducted particular analyses and to memorialize what information was developed. Analysts should enter data into laboratory notebooks in a way that minimizes the possibility that data can be covertly altered after the analysis has been done. Standard practices for all entries into the notebooks should require that: (1) all entries should be made with non-erasable ink; (2) all deletions, additions, and corrections should be made with a single line through the data and initialed and dated by the person making the change; (3) the reason for any change should be described in the notebook; (4) all blank spaces should be eliminated by drawing a diagonal line through such areas; (5) no entries that are irrelevant to the analysis should be recorded in the notebook; and (6) the analyst should sign the notebook at the end of each area where analytical data are recorded.

The physical aspects of keeping the notebook and the recording of information are very important in maintaining laboratory notebooks, and the laboratory should take all reasonable measures to reduce the possibility of altering the notebooks after an analysis. Although a common practice, laboratories that are responsible for compliance-related analyses should avoid the use of bench sheets from which data are later transcribed into notebooks. Instead, laboratories should enter the data directly into bound notebooks in which the page numbers have been consecutively printed. Most commercial printers can prepare such notebooks from an exemplar format furnished by the laboratory. The purpose of using bound notebooks is ensure that pages have neither been removed nor inserted into the book after the analytical work has been done.

The process of maintaining a laboratory notebook does not end with the completion of an analysis. Once full, laboratory notebooks should be stored in a secure location to prevent theft or access by unauthorized persons. In addition, laboratory supervisors should review each notebook weekly to ensure that the entry procedures are being followed. The entries should be inspected for signs that may suggest that an analyst misunderstands the procedures or is acting improperly. The signs that improper conduct may have occurred include unusually similar data, unusually variable data, data that are inconsistent with the analytical procedure, missing entries, repetitive entries of the same number where this is not expected, and numbers outside the expected range. After reviewing a notebook, the supervisor should sign and date the notebook at the end of the reviewed data to indicate that it has been inspected.

Another important element of a laboratory compliance program is establishing a policy for dealing with problematic results. Although reasonable efforts should be made to avoid problems in the laboratory, it is essential to prepare for such possibilities, as errors of judgment made after a problem is discovered can have grave consequences. When problems occur, laboratories should be very careful in how they address the associated regulatory issues. First, complete honesty is essential, as discovery of a coverup or false reporting on the part or the laboratory will probably result in a criminal investigation. Second, because prompt and voluntary disclosures of violations are often the best means of avoiding enforcement, laboratories should inform the party requesting the analysis as soon as problems are discovered. This practice not only provides the party with more control over how to disclose the problem, but, in some instances, it may also allow for resampling and avoid the compliance issues altogether. Even when resampling is not an option, the time can be used to determine the best way to address the regulatory issues. In situations such as these, an impeccable reputation for honesty and sensitivity to the needs of regulatory agencies is a very valuable asset. A reputation for cutting corners, concealing problems, and evading responsibility will usually compound the difficulties; therefore, candor and timeliness are a better basis for the long-term management of laboratory issues.

9.2.3 Contracting for Analytical and Monitoring Services

Most regulated parties and many consulting firms contract all but the most elementary analytical work to independent laboratories. Similarly, some monitoring services such as the stack testing of air emission sources are mostly contracted out. Unfortunately, many organizations fail to consider the complicated issues of environmental regulation and commercial law associated with these transactions. Unless such issues are considered when procuring analytical services, laboratories are placed in situations where they can easily fail to provide results that meet the needs of their customers. One consequence is the failure to meet environmental requirements, as regulated parties remain liable for analytical or monitoring problems caused by outside contractors. To protect themselves, organizations should adopt good contracting practices when procuring analytical and monitoring services. In addition, such practices will help protect organizations from laboratories that provide incorrect or otherwise deficient services.

The first step in procuring laboratory services is to find a laboratory that is capable and willing to do the work needed to meet a compliance requirement. It can be difficult for outside parties to assess the competence of laboratories, but a starting point is to ask for a statement of qualifications. The laboratory can also be evaluated by judging the willingness of the laboratory director to discuss the capabilities of the laboratory and its attitude toward meeting compliance needs. Recommendations from other customers are another source of information about laboratories, although such opinions may be meaningless unless the customers are sophisticated enough to evaluate the quality of the services provided. Another approach to assessing laboratories is to send commercially prepared performance evaluation samples to serve as a check of their capabilities and ability to comply with instructions, such as requests that the laboratory use a specific analytical method.

Parties contracting with laboratories should generally emphasize communicating what services they need rather than focusing on legal issues such as indemnities and disclaimers of responsibility. Although reaching agreement on the consequences of improperly performed services can be important, the fact that a laboratory messes up an analysis will not excuse noncompliance by the regulated party. By ensuring that the laboratory understands what is expected by the customer, the chance of such noncompliance can be substantially reduced. Putting the appropriate level of effort into communicating specific analytical and regulatory needs will also go a long way toward reducing the potential for disputes after the services have been rendered.

The basic information that should be communicated to the laboratory includes the number of samples to be delivered, the matrices to be analyzed, and the parameters to be analyzed. The customer should also identify the purpose for which the analytical data will be used and any other special requirements. In some cases, it may also be necessary to specify detection limits, quality control requirements, report format, and deadlines for providing results. In response to this information, the laboratory should either confirm that it can meet the requirements or decline to accept the work.

A major item to address in the contract is whether the permit or regulations specify that particular analytical methods be used. For example, the EPA requires data in permit applications and discharge monitoring permits submitted pursuant to the NPDES program to be determined using test methods approved for that program. Similarly, the EPA has approved another set of test methods associated with the Resource Conservation and Recovery Act (RCRA). Generally speaking, methods approved for specific programs cannot be used interchangeably. Unless laboratories are told that the data will be used for regulatory reporting, the customer will have no assurance that proper test methods will be used.

Another issue that may need to be discussed with the laboratory is detection limit requirements. Even laboratories using the same types of equipment and test methods can report different detection limits depending on their quality control procedures; therefore, it is a good practice to ascertain whether the laboratory can achieve the desired detection limits. Most regulatory programs, however, do not specify detection limits, and in such cases organizations have little to gain by specifying detection limits. For example, the typical NPDES permit requires the use of methods approved in 40 C.F.R. Part 136 but does not otherwise mandate detection limits. Organizations that specify detection limits when contracting for NPDES analyses may spend more money to achieve enhanced but unnecessary analytical performance but gain no compliance benefits.

As noted previously, many environmental professionals view laboratories merely as providers of standardized services and fail to appreciate them as technical resources. For instance, many laboratories can provide valuable information about possible impediments to analyzing samples if the customer takes the time to describe what is being submitted. Likewise, taking the time to explain the purposes and associated regulatory requirements that underlie a request for analytical services can alert laboratories to sampling issues and opportunities to reduce costs. Consulting with laboratories rather than merely placing orders can be an inexpensive means of avoiding compliance problems and controlling costs.

It is highly recommended that arrangements for analytical services be made well in advance of sample collection dates. Many analytical methods have prescribed holding times which limit the amount of time that can elapse between the time the sample is collected and the time it is analyzed. Once the samples are collected, customers will have diminished negotiating leverage in procuring laboratory services as there will be less time to contract with another laboratory should the intended laboratory be overloaded or demand unreasonable terms. Also, it is in the customer's interest to ensure that the laboratory is capable of doing the analysis and has sufficient time in which to prepare and analyze the samples.

As stated above, parties should focus first on communicating what they want from the laboratory or testing contractor. This does not mean, however, that legal issues should be avoided. Contracting for these services often involves a formal contract with provisions governing indemnities and disclaimers. Whether parties are willing to accept such terms will depend on factors such as tolerance of risk, the desire for the services, and insurance coverage. For instance, a laboratory may attempt to disclaim liability for consequential damages should its negligence result in the loss of a sample or a defective analysis. What this means is that the laboratory will not compensate the customer for any loss other than the cost of performing the analysis such as the cost of collecting new samples. Customers who will have incurred substantial expenditures in collecting the samples or will be exposed to significant penalties for failing to meet monitoring requirements should not take this issue lightly. Should a laboratory insist on disclaiming this liability, the customer will have the option of accepting the terms or finding another laboratory to perform the service. Provided the customer begins the procurement process early enough, it can generally find a laboratory that will accept full responsibility for the actions within its control.

Many times, the parties to a laboratory contract fail to discuss who will bear the loss in the event that the services do not meet the expectations of the customer. In such cases, the law will generally assign responsibility for damages to the party in the best position to prevent the loss. For example, the responsible party can be a customer who fails to identify the purpose of the analysis or a laboratory that fails to follow explicit instructions. Because analytical service contracts seldom lead to legal disputes, it is not unusual for services to be ordered with minimal attention as to who assumes the risk of loss. Unless the parties are comfortable in leaving the issue of who bears the risk of loss to general contract law, they should reach an understanding of the purposes for which the data are needed and address who will be responsible for losses.

Organizations should carefully review the test results to see if they conform to the terms of the contract and meet compliance needs. Specific items that should be checked are whether the laboratory used approved methods, applied the correct quality control requirements, and met holding time requirements. The failure to check results can allow deficient performance to continue indefinitely and may also weaken the ability to assert legal claims because of legal defenses such as waiver and statutes of limitation. If problems are found, the first step should be to confer with the laboratory or testing contractor and give it an opportunity to correct the deficiencies. However, issues such as short holding times or limited samples will in many cases preclude corrective measures. Depending on the circumstances, it may be appropriate for the laboratory to repeat analyses, waive charges, or compensate the customer for the costs of collecting new samples.

Even when corrective action is impossible, issues regarding deficient work should be handled tactfully and respectfully whenever possible, as the laboratory's cooperation may be needed later to help resolve regulatory difficulties. In egregious cases, litigation may be warranted to obtain compensation for damages caused by a laboratory's failure to deliver adequate services. Whether such actions will succeed will largely depend on how well the contract describes the services needed and the respective obligations of the parties. Furthermore, when the parties have or desire a longstanding business relationship, they should consider alternatives to litigation such as mediation. However, it cannot be overemphasized that diligent contracting practicing is a far more effective and efficient way to deal with analytical issues than trying to resolve disputes after the fact.

9.3 ESTIMATING AND REPORTING ENVIRONMENTAL QUANTITIES

Environmental estimates necessarily involve the concepts of uncertainty and applicability. Unless these concepts are understood, the process of estimating and reporting environmental quantities can present serious compliance risks. To ensure environmental compliance it is important to understand the limitations of most quantitative estimates and how they affect compliance matters.

Measuring quantities in most environmental contexts involves some degree of estimation and therefore the exercise of judgment. The use of judgment in developing estimates also introduces the tendency to skew the result high or low depending on the circumstances, but even the introduction of conservatism into environmental estimates is not without risk. Estimates that are conservative in one context may be too lenient in others. For example, providing estimates in permit applications of air emissions that are intended to reflect well-operated emission control technology are conservative in the sense that such estimates will likely result in stricter emission limits. Using low estimates, though, may cause a facility to avoid new source review if they suggest that emissions will fall below the threshold at which review is required. In such a case, should the actual emissions exceed the estimates, then the operator of the facility could find itself subject to enforcement actions. To avoid charges of avoiding new source review, some organizations may be tempted to skew their estimates upwards, but using higher estimates may result in the agency placing unreasonably lenient emission limits in the permit. Because these kinds of limits will reduce the burden to control or limit actual emissions, they can later expose an organization to charges that it misled the government to avoid more stringent pollution control.

Government agencies have a tendency to complicate environmental estimating by failing to consider the inherent uncertainty of estimated quantities when promulgating regulations. This can make compliance determinations difficult when a regulatory requirement is triggered at a quantitative threshold, but estimating the quantities at a reasonable degree of precision is impossible. For example, many of the widely available factors for estimating air emissions were developed from data that are woefully outdated, and data from newer sources are often insufficient to develop reliable estimates. In some instances, government agencies have issued guidance on how to deal with such issues. For example, the EPA's *Technical Support Document for Water Quality-Based Toxics Control* provides detailed information regarding how the agency accounts for effluent variability when determining effluent limits.[77] In most cases, though, environmental professionals will have to apply their individual judgment in determining how to report quantities.

The better approach to dealing with the problem of how to report quantities is to identify and account for the uncertainties when estimating and reporting environmental quantities. Issues that should be noted include an overview of the data used to develop estimates, the sensitivity of the estimated quantities to fluctuations in the underlying data and assumptions, and the potential regulatory consequences associated with the approach to developing an estimate. By using sound estimating practices and qualifying information when appropriate, environmental professionals can do much to avoid serious compliance problems.

A good example of the issues involved in determining environmental quantities can be found in the air pollution control field. Because air emissions tend to be difficult or expensive to measure continually, most emission sources are tested intermittently. To overcome the lack of continuous monitoring data, it is usually necessary to rely on air emission factors to estimate the pollutants released by activities and operations. Unfortunately, many commonly used emission factors are outdated and statistically unsupportable. In fact, most emission factors published in publicly available reference sources are based on a few data from testing conducted during the 1960s and 1970s when control technology and practices were less effective at limiting emissions. Furthermore, the limited availability of test data in prior decades precluded meaningful statistical analysis of the utility of these factors in predicting actual emissions. Even though these emission factors are unlikely to reflect emissions from modern sources, they are still widely used.

The traditional approach by which most emission factors have been developed has consisted of: (1) collecting emissions data and associated operating parameters such as production or fuel consumption rates, (2) adjusting the data by dividing the emissions data by a selected operating parameter, and (3) computing the mean of the adjusted data to obtain an emission factor for the process or activity. Emission factors developed using this approach are nothing more than simple linear models and do not account for phenomena such as boiling points, reaction thresholds, and nonlinear effects. Furthermore, the traditional approach to developing emission factors rarely incorporates statistical evaluation to assess whether emissions actually correlate to their associated parameters (e.g., production rates). In addition, simple models do not account for the variability and quantitative uncertainty associated with most emissions.

Although statistical analysis is rarely used to develop emission factors, the increased availability of test data and software for statistical analysis makes it relatively easy to develop emission factors that are more reliable than those set forth in the standard references. In addition, using statistical techniques such as regression analyses allows parties to develop factors that can account for uncertainties. One approach to developing better emission factors is to record all possibly relevant data during testing such as production rates, operating temperatures, and raw material usage. This will allow the use of multivariate models that can evaluate more comprehensively the effect of operating parameters on emissions. These models can also provide better information about emission variability than simpler models and allow for the establishment of confidence limits for the estimates. Such models can provide industrial sources with a powerful tool by which to increase the flexibility of operating permits. For instance, by establishing that emissions vary in response to multiple production parameters, sources can avoid operating restrictions that depend on single parameters such as production rates. Similarly, industries may be able to discern and verify which parameters constitute the best surrogate indicators for ensuring compliance with emission limits.

Case Study: Statistical Analysis of an AP-42 Emission Factor for Granular Triple Superphosphate Manufacturing

The U.S. Environmental Protection Agency publishes a reference of emission factors known formally as the *Compilation of Air Pollutant Emission Factors* but more commonly referred to as AP-42.[85] Although the EPA strongly cautions in the introduction to AP-42 that the factors have substantial limitations, they are nonetheless widely used to develop estimates used for permitting, air emission inventories, and as the basis for emission fees. Most of the emission factors in AP-42 are presented without statistical information that would indicate their reliability. Instead, AP-42 rates the factors on a scale of A to E based primarily on the number of data used to develop the factors.

An evaluation of one of the factors in AP-42 confirms that they can be very unreliable. For example, triple superphosphate, commonly known as TSP, is used as an ingredient in fertilizers and also as a cleaning agent. Operations that manufacture granular triple superphosphate (GTSP) can emit significant amounts of fluoride. To assist in the estimation of fluoride emissions from these operations, AP-42 provides an A-rated factor that purports to describe representative emissions of fluoride based on the rate at which a GTSP plant utilizes phosphoric pentoxide (P_2O_5) in the manufacturing process.

According to AP-42, emissions from the reactor, granulator, dryer, and screening sequence should emit fluoride at a rate of 0.12 kilograms per metric ton of phosphoric pentoxide processed. This emission factor was computed by taking the mean of 25 production-adjusted emission tests from seven facilities.[90] Because the emission data from which the factor was developed are still available, it is possible to evaluate the statistical association between the observed emission rates of fluoride and the rate at which phosphoric pentoxide was used in the manufacturing process. The data were fitted to several simple regression equations including linear, exponential, power, and quadratic equations. The AP-42 factor, best fit nonlinear equation, and best fit linear equation are illustrated in Figures 9.1, 9.2, and 9.3. All the figures show that fluoride emissions correlate poorly with the phosphoric pentoxide feed rate. Not surprisingly, the emission factor correlates as poorly or worse than the best-fit equations. In other words, statistical evaluation of the emission data demonstrates that the phosphoric pentoxide rate cannot accurately predict or estimate fluoride emissions from these

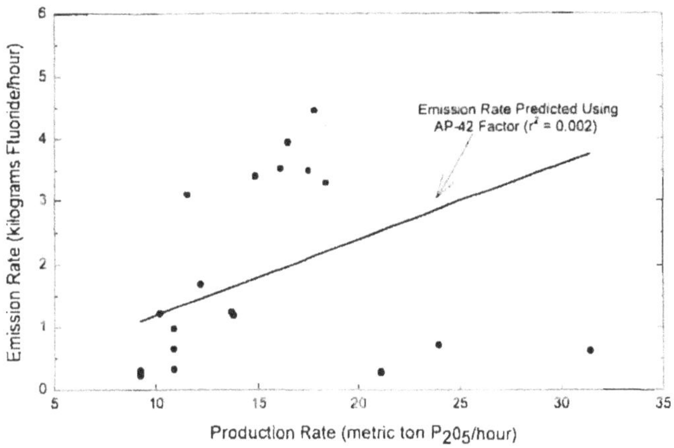

FIGURE 9.1
AP-42 emission factor and fluoride emission data collected from reactor, granulator, dryer, and screening stacks at GTSP production plants.

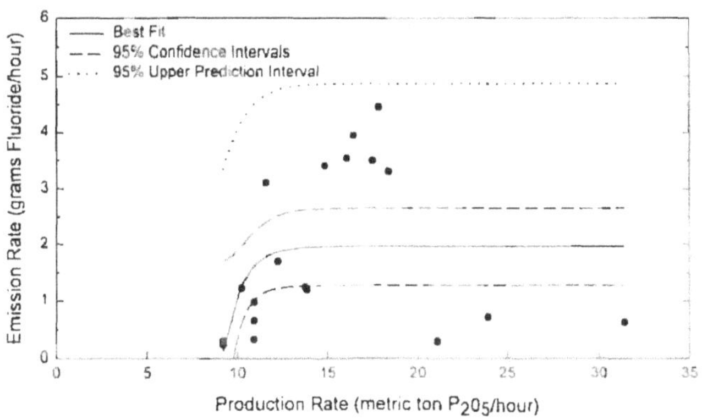

FIGURE 9.2
Best exponential fit of fluoride emission data collected from reactor, granulator, dryer, and screening stacks at GTSP production plants.

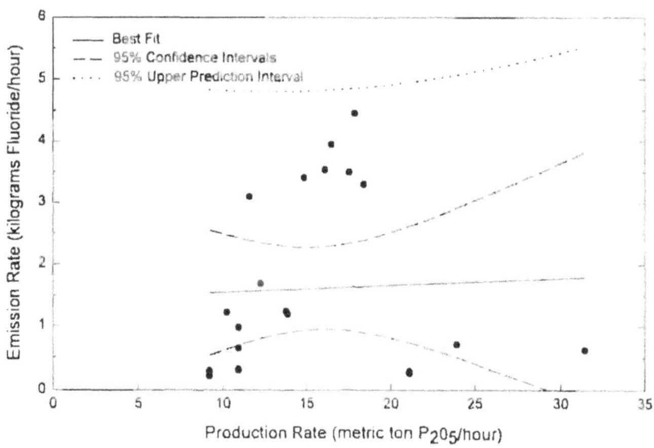

FIGURE 9.3
Best linear fit of fluoride emission data collected from reactor, granulator, dryer, and screening stacks at GTSP production plants.

sources. This means that the AP-42 factor, despite its A rating, is scientifically invalid as an emission factor.

Naturally, the use of invalid emission factors is not without the potential for regulatory consequences. Using the factor described in AP-42 to estimate emissions from the seven GTSP plants from which the emission data were collected indicates that fluoride emissions for most of the plants would have been underestimated or overestimated by at least a factor of 2. In fact, this emission factor produces a reasonably close estimate for only one facility. The disparity between the predicted and actual emissions for some facilities is great enough to suggest that reliance on any of the production-based emission models could cause significant problems if used in contexts which have compliance issues.

9.4 SURROGATE PARAMETERS

Operating parameters such as production rates, temperatures, and ingredient concentrations are often used as regulatory surrogates, such as when a production rate is restricted in an attempt to limit air emissions. While the use of such parameters can simplify measurement and control issues, environmental professionals must be careful to ensure they are used only when suitable for the intended application. Among the issues to consider when using operating parameters as surrogates in regulatory settings are their relevance to the phenomenon being controlled and their ability to be measured accurately.

The relevance of an operating parameter will vary according to how it is used, but in any case only parameters that adequately relate to the regulatory activity should be used. For example, production rates of manufacturing operations can be expressed in terms of raw materials, intermediate products, or final product. Generally speaking, using the overall production rate as regulatory surrogates should be avoided for complex processes, as emissions and discharges tend to correlate better with specific unit operations. One way to illustrate this principle is to consider how an operating parameter could be selected to limit arsenic emissions from a copper smelter that occasionally processes ores containing high concentrations of arsenic. In this case, the best parameter would probably be the quantity of arsenic in the ore, as arsenic emissions are more likely to correlate with the amount of arsenic entering the process than the amount of copper leaving the process. Using the copper production rate as a regulatory surrogate would ignore periods when high-arsenic ores are not being processed and could result in overly stringent production limits. Moreover, incorporating the arsenic content into the surrogate would give the smelter more flexibility in planning its operations because it would be free to adjust the feedstock consumption to comply with production limits.

The ability to measure operating parameters needs to be considered before selecting them as surrogates because many parameters can be relevant to a regulated activity but very difficult to measure accurately. Particular attention should be given to whether variability will impair the utility of measurements and how parameters will be measured. Changes in density can affect whether a parameter should be measured in terms of mass or volume, as can practical considerations such as the availability of measurement devices. For example, materials such as liquids are easy to measure with equipment such as flow meters. Other transfers, such as the transport of low-density solids on conveyor belts, can be difficult to measure accurately. The measurement point also affects the suitability of an operating parameter as a surrogate. Ideally, operating parameters should be measured at points where they are most likely to reflect the regulated activity. Because parameters such as flow rates and temperatures may change throughout a system, measurements should normally be made at points that are close to the regulated activity. This is particularly the case when materials temporarily accumulate at various points throughout a system.

Finally, regulated parties should consider how monitoring operating parameters fits in with nonenvironmental activities such as production recordkeeping and accounting. It is not unusual for operations to maintain records such as monitoring production efficiency or accounting, but the degree of accuracy and precision required for these purposes is often considerably less than what is needed to satisfy regulatory requirements. In addition, the time frames of these measurements may not coincide with those required by regulations or permits, thus it is important not to assume that internal records are suitable for compliance-related activities. Accounting records sometimes seem to provide useful information but on further examination are found to be inadequate. Accountants use generally accepted accounting principles, which often allow for considerable latitude when developing estimates used for internal accounting purposes. For instance, an accountant may assume that byproducts are produced at rates proportional to the primary product and not use direct measurements. Such an approach may provide an organization's management with adequate financial information but may be entirely insufficient when used for compliance purposes.

10 Compliance Assistance

Outside resources such as consultants, environmental service contractors, and attorneys can provide valuable assistance to organizations faced with compliance issues. Furthermore, the economics associated with environmental compliance require most organizations to contract for some level of environmental services. Used properly, these services can facilitate compliance at lower cost and higher effectiveness than can be achieved by using internal resources; however, contracting for services requires that the organization surrender some control of how activities are conducted. Delegation poses no problems when the activities are performed competently and in accordance with the client's principles, but indifferent or inadequate services can create situations in which noncompliance becomes inevitable.

Sensitivity to compliance issues varies among technical consultants, environmental services providers, and attorneys. While many place great importance on environmental compliance, some are indifferent. A brazen few seem to take perverse pride in their professed ability to violate environmental laws and elude discovery. In many cases, violations committed by outside parties will be attributable to the organizations that hired them; therefore, it is important to consider whether such parties are willing and able to comply with environmental laws before retaining their services. In addition, organizations should make clear that the outside parties who work for them are expected to abide by the organization's policy of total environmental compliance. This message should be frequently reinforced through contract provisions, specifications, routine communications, and the actions of the organization. While organizations need not be unduly repetitious, it is important that no misunderstandings develop regarding their commitment to compliance.

10.1 TECHNICAL CONSULTANTS

Organizations rely on technical consultants to provide particular expertise that the organization does not have or to perform work when it would not be economical to hire the employees that would otherwise be needed. To use consultants effectively, clients should understand that the nature of the project will in large part dictate the type of firm and the qualifications of the individuals that will be needed to perform the work. For instance, some projects demand services from people who possess the highest levels of technical proficiency, while others require consultants who can perform more mundane tasks with maximum timeliness and efficiency.

Not all environmental services directly involve compliance issues, although most have the potential to affect compliance. For example, organizations are not legally required to perform environmental audits to assess potential liabilities prior to purchasing properties although this is commonly done. Nevertheless, it is possible that an audit could disclose noncompliant conditions at the property for which the law requires action. Clients should therefore not only evaluate the suitability of prospective consultants for the work, but also consider how the work could affect compliance.

A consultant's attitude towards compliance issues is an important consideration for entities that seek total environmental compliance. Some consultants are naturally inclined to consider compliance issues when working for clients, but many are not. A few consultants even exhibit dangerous attitudes in which they actively condone or even encourage clients to engage in unlawful conduct; however, many consultants fail to give compliance issues their due consideration simply because such concerns do not directly affect the completion of their projects.

Another reason for concern is that, because consultants are independent contractors, they often face a lesser degree of risk with respect to compliance matters. For instance, when a consultant incorrectly estimates the air emissions from a client's facility, it is the client who faces the risk of being charged with violations and not the consultant. Without direct responsibility for the environmental violations committed by clients, some consultants may not be sufficiently sensitive to compliance issues to avoid problems. Clients need to assess the sensitivity of consultants toward compliance before selecting them if problems are to be avoided; otherwise, the consultant may not have considered the effort and time that compliance issues require when seeking the work and may be inclined to dismiss their importance. In fact, it is not unusual for consultants to try to persuade clients that they can disregard unexpected or cumbersome requirements that impede the completion of their projects. Although such attempts should be rejected, a better approach to dealing with these situations is to encourage the consultant to consider the possibility of compliance issues at the inception of the project.

The most straightforward way to encourage consultants to consider the importance of compliance issues is to let them know how important such issues are to the client. Few consultants dismiss such concerns, and most are willing to accommodate them when alerted early enough. In cases where consultants downplay concerns about compliance, the client is better off not using their services. Such consultants can only create problems, as their work will not reflect sound compliance practices, and they often will use their position as professed experts to persuade the organization's employees that environmental compliance is unimportant. In addition, when enforcement matters arise, such consultants tend to forget or deny that they disregarded the compliance aspects associated with the work they did for the client. This is a double disservice to clients, as the subsequent lack of cooperation makes it more difficult to resolve enforcement matters.

Another thing that can be done to promote compliance is to assist the consultant in dealing with compliance issues. Many organizations assume that consultants will generally understand the compliance issues associated with their projects, but this is not always the case. Unless the issue of the contractor's knowledge is addressed, it is even possible for consultants to violate environmental laws based on a good-faith misunderstanding of the requirements that apply to the project. Although the client should not assume the legal responsibility to inform a consultant of the applicable legal requirements, it does not hurt to verify that consultants are aware of applicable requirements or to provide relevant information about them. For example, consultants should be provided with copies of permits that may be relevant to the consultant's activities. Likewise, it is a good practice to inform consultants of lesser known requirements such as local ordinances.

Consultants' ethics are another area to evaluate, and it is extremely important that consultants be honest. Although most consultants in the environmental field conduct their practices with the highest levels of integrity, there are a few whose ethics permit them to practice with some degree of deceit. Particularly troublesome are instances when a lack of honesty can cause clients to face enforcement problems which include reporting and disclosure obligations. In general, any consultant who feels that clients should seize opportunities to evade regulatory requirements has the potential to cause compliance problems. Even if enforcement does not materialize, honesty generally correlates with credibility which is of great value when consultants are required to deal with agencies. Concerns about a consultant's integrity can lead to protracted and more expensive regulatory activities such as permitting matters. In addition, consultants' ethics are often attributed to the clients, thus retaining firms or persons of questionable integrity can result in agencies viewing the client with higher levels of suspicion.

Although most consultants are honest, many could do a better job of respecting ethical principles such as confidentiality or loyalty. For instance, consultants should not disclose sensitive information about their clients to the clients' competitors or potentially adverse parties without the knowledge and consent of the client. Unfortunately, many consultants do so in routine conversations and while marketing their services. If your organization has retained a consulting firm that provides you with private information gleaned from their other clients, you can be reasonably sure that the consultant will reveal similar information about your organization to others. When evaluating or retaining consultants, it is a good practice to state that confidentiality is expected of consultants. The better consultants will respect the client's wishes in this regard, and the others will not be able to claim that they did not realize that confidentiality was important to the client.

Another ethical consideration is the willingness of consultants to communicate with their clients. At a minimum, clients should know what tasks their consultants are working on and the expected costs. Also, clients are entitled to have their telephone calls answered or returned promptly. Once again, most consultants do a good job of communicating with their clients, but working with a consultant that does not feel obligated to communicate with clients is frustrating from a business perspective and also presents compliance risks. The client ultimately bears the responsibility for environmental problems regardless of a consultant's role in causing them. Prompt communication of important issues and developments allows clients to act responsibly and effectively regarding potential compliance issues.

It is obvious that sensitivity to compliance issues and overall integrity must be evaluated before a consultant is selected. Good consultants are forthright about what they can do for their clients and do not seek work for which they are not qualified. Unfortunately, there are consultants who freely overreach in marketing their services and will say almost anything to get business. When this happens, many clients get saddled with expensive and improperly executed projects because the consultants lack the expertise needed to perform the services proficiently. Not only does such performance lead to compliance problems, but the projects are also sometimes capitalized on by the consultant to get additional work. Some of the signs of overreaching marketing practices are an inordinate number of marketing visits, frequent allusions to "necessary" future projects, and representations that the consulting firm is uniquely qualified to provide the services being offered to the client.

Sometimes consultants may overstate compliance issues in an attempt to make the services appear mandatory or at least more imperative. This can induce clients to devote management and compliance resources to areas where they are not needed, with the result that insufficient attention is paid elsewhere. Another problem common to overreaching consultants is that they are often inclined to ignore or take risks with respect to compliance requirements. If the services push the client into unnecessarily complex solutions, the client becomes needlessly exposed to compliance problems, such as having to operate and maintain equipment that is less reliable than would otherwise be required.

Another potential problem with consultants who are obsessed with marketing is that concerns about alienating prospective clients cause them to fail their existing clients at inopportune moments. For example, a consultant may tailor reports to avoid conclusions or present data that may adversely affect other clients. Similarly, a consultant who refuses to support a client when a matter becomes litigated for fear of offending other litigants will at best be useless and in some cases an outright liability. Although they may be willing to attend depositions or testify in hearings (often for an augmented fee), they may avoid making definitive statements or claim memory lapses with respect to critical facts. When this happens, the client will be forced to find other witnesses to fill the void. Furthermore, diffident testimony by a party's consultant tends to create negative inferences that weaken their legal positions.

Although attitude and integrity are prominent compliance concerns when working with consultants, competence is also important, as poorly performed services can lead to compliance problems. The fundamental requirement of a consultant is mastery of the technical principles required to carry out their assigned tasks. Because there are few regulatory barriers that prevent unqualified persons from becoming environmental consultants, organizations should determine whether prospective consultants are, in fact, qualified to perform the desired services. At a minimum, organizations should assess a prospective consultant's educational background, work experience, and professional registrations. Better yet, organizations can check references and consult with persons who are familiar with a consultant's professional reputation.

Another aspect to consider when selecting consultants is their approach to providing services. Some consultants start each project afresh and thoughtfully incorporate processes and procedures that ensure the finished project is founded on sound scientific and technical principles. Others may blindly follow their routine procedures without considering whether the results reflect reality. These consultants present special compliance risks when their judgments will be evaluated according to a standard of perfect hindsight. For example, a consultant who estimates air emissions using factors developed 30 years ago without considering better and more recent information may expose the client to serious enforcement actions if the estimates are inconsistent with the actual emissions. Unfortunately, it is the client that must deal with the consequences of using consultants whose level of proficiency is insufficient to meet the associated regulatory demands.

Another area where caution is warranted is using technical consultants for purposes for which they are unqualified. Many people, including some consultants, believe that the ability to understand issues associated with other professions qualifies them to advise clients on such matters; however, using technical consultants as advisors on business or legal issues is an invitation to trouble, although many consultants relish these roles. While consultants can sometimes provide valuable information with regard to business and legal matters, they are not sufficiently qualified in these areas to render competent advice. For example, even though a consultant may be familiar with regulatory requirements, they do not have the education or experience regarding litigation issues, administrative law, and legal construction that would qualify them as legal advisors. Similarly, consultants generally do not have sufficient knowledge to advise clients on business decisions, although they are sometimes positioned to provide factual information to decision-makers. Unfortunately, while it is obvious to most managers that bad business advice from consultants may result in poor financial performance, they are not as alert to the fact that bad legal advice will not preclude adverse legal consequences. As some companies have found, however, bad legal advice from consultants can result in stressful and painful enforcement experiences.

Case Study: *United States v. White*, 766 F. Supp. 873 (E.D. Wash. 1991)

The Facts

One of the most sensitive determinations in environmental law is deciding when materials that would otherwise constitute hazardous waste are not regulated under the Resource Conservation and Recovery Act (RCRA). There are several exemptions from such regulations, including a recycling exclusion, but applying them to specific situations often requires expert legal interpretation. One company decided to rely on the recycling exclusion to dispose of pesticide wastes after seeking advice from a consulting firm but ended up facing criminal charges.

In the early 1980s, PureGro, Inc., installed an evaporator tank at its Pasco, WA, facility to reduce the volume of pesticide rinseates produced from washing out equipment and containers. The tank was used to accumulate rinseate for about 4 years, at which time PureGro contacted a consulting firm in regard to what to do with the material. The firm analyzed the accumulated rinseate and advised PureGro that it would not have to handle the material as hazardous waste if PureGro found an application for the material consistent with its use as a pesticide. However, PureGro was further advised that if the material was declared a waste, it would have to be disposed of as hazardous waste.

Acting on this advice, PureGro loaded the rinseate onto a truck and sprayed it on a field. Following an investigation by state and federal authorities, PureGro and four of its employees were indicted for violating the RCRA and the Federal Insecticide, Fungicide, and Rodenticide Act (FIFRA).

Discussion

A dominant issue in the case was whether the vagueness associated with the recycling exclusions precluded prosecuting the defendants on the grounds of due process. The Court ruled that the RCRA recycling provisions, although detailed and complex, were not so vague as to preclude the prosecutions. Instead, the Court framed the ultimate issue in the case to be whether the defendants had accumulated the rinseate for the purpose of beneficial use. Because this issue required a factual determination as to the defendant's intent, the Court ruled that the jury would need to decide why the company was accumulating the rinseate. The fact that the defendants had been advised by a consulting firm that the material was not regulated under the RCRA if beneficially reused was not relevant to the Court's decision, as the good faith reliance on the advice of a consultant is not a legal defense.

Although the Court ruled that the jury must decide the factual issues, the matter was never presented to a jury. PureGro agreed before trial to plead guilty to the FIFRA count, and the government agreed to dismiss the remaining charges against the company and all the charges against the individual employees. Later, the U.S. Department of Justice was vociferously criticized by some interests within the EPA, environmental groups, and some members of Congress for dismissing the charges against the individuals. In a stunning departure from previous practice, Attorney General Janet Reno agreed to require the career prosecutors associated with this case to testify about their decisions in a politically charged hearing before a House subcommittee. Although it is the official policy of the Department of Justice not to allow political considerations to influence the exercise of prosecutorial discretion, plea agreements in which companies accept responsibility for criminal conduct in exchange for the dismissal of charges against employees became less common after the PureGro case was resolved.

10.2 ENVIRONMENTAL SERVICE PROVIDERS

Many organizations need to contract for environmental services such as hazardous waste disposal, environmental testing, and equipment maintenance. Most of the providers of such services are honest, although the potential to encounter unscrupulous operators seems to be higher in certain fields such as waste disposal. Fortunately, there are several efficient ways to evaluate service providers before hiring them. In addition to requesting references, a reasonably effective way to evaluate potential providers is to obtain credit histories from outfits such as Dun & Bradstreet. Credit reports can provide information such as credit problems and judgments that might indicate potential unethical practices or financial impediments to their ability to perform. Regulatory agencies can be another good source of information about service providers. Inquiries to agencies can determine whether a service provider has all the necessary permits and registrations to conduct the proffered services lawfully. In addition, agencies can provide information about compliance histories and whether the provider has been involved in incidents such as spills or other emergencies. Some agencies maintain databases of this type of information that can be viewed via the Internet.

Of course, prospective service providers should be queried regarding their experience and qualifications. Experience is very important, as providers who provide services in new areas are sometimes unaware of regulatory requirements or are unable to meet them. For example, general demolition contractors may not know all the requirements associated with asbestos abatement and probably would not have the equipment to perform such services lawfully. Signs that a service provider may be unqualified for a particular project include unrealistically low prices or bids, references solely from customers who procured services substantially different from the one being sought, and flippant or unbefitting sentiments toward regulatory requirements. Requests to borrow routine protective gear or questions that indicate the provider is unfamiliar with the relevant regulatory requirements also indicate that a provider may be unqualified to render the proffered services.

Aside from independent references and objective qualifications, it is prudent to evaluate the attitude of the service provider. A particular area of concern is the willingness of service providers to be held accountable for their misconduct or negligence. One way that some service providers attempt to avoid accountability is to insist on contract provisions which attempt to waive liability for acts solely within the provider's control. In more egregious cases, the provider's contract will require that the customer indemnify the provider against liability even when the problem is solely the fault of the provider. Whether such provisions indicate that the provider is overreaching or is genuinely prone to negligent or delinquent conduct is, from a practical perspective, irrelevant because the market for environmental services has matured to the point where providers lack the leverage to force customer to agree to waivers and indemnities.

Another indication of an attitude problem is an unwillingness to provide proof of a reasonable degree of insurance coverage. Failure to carry adequate insurance may indicate problems such as business naiveté, financial difficulties, or a poor claims history. As with the case when contract provisions disclaim accountability, the reasons for not carrying adequate insurance may not have practical relevance, as there is seldom adequate justification for not carrying a reasonable amount of insurance. Using providers who may not be able to defend or satisfy legal claims made against them is poor judgment because it invites claims against the organizations that hire them.

10.3 ATTORNEYS

Environmental law is a complex specialty, and organizations should seek counsel on environmental matters from attorneys who specialize in environmental law. One reason to use specialists is that environmental matters are often regulated in ways that evade superficial rationality. Attorneys who rely on the uninformed application of commonsense principles to environmental legal analyses can thus commit grave errors. Furthermore, many legal principles such as causation, equity, and the duty of disclosure can apply to environmental matters differently than they do in other legal contexts.

Although environmental attorneys must be able to cope with the tremendous volumes of legal authorities that constitute environmental law, it is also essential that they be able to comprehend technical issues. Most environmental attorneys have little or no technical education and therefore do not have strong technical credentials. Nonetheless, many environmental attorneys are able to understand the technical concepts involved in environmental matters and provide competent service to their clients. Clients should also be aware that attorneys who are not proficient at understanding technical issues may be inclined to let them down in several ways. For instance, attorneys who struggle with the technical aspects of a matter may miss some of the legal ramifications, as the applicability of many environmental requirements depends on technical considerations. Also, attorneys who rely extensively on outside advisors, such as technical consultants, may tend to unreasonably defer to them regarding the legal aspects of a matter. In some cases, attorneys that have difficulty understanding technical issues may encourage their client to accept less favorable settlements because the attorneys either cannot competently assess the client's position or are uncomfortable pressing forward with a matter they do not understand.

Another consideration is the level of service that a client can expect from an attorney. Environmental issues often require regulated parties to act promptly and decisively, and they need to obtain timely and definitive advice when problems arise. For example, clients who have spilled hazardous substances cannot wait several days to have a telephone call returned if they are to ensure compliance with reporting and response requirements. In general, clients should expect attorneys to return their telephone calls promptly and pay diligent attention to the matters with which they have been entrusted. Also, clients should expect their attorneys to provide definitive answers to their questions even if those answers must be qualified to reflect the uncertainty that permeates environmental law. Although attorneys tend to be very busy and some allowances must be made, clients do not have to tolerate tardy communication, procrastination, or oblique responses to their queries.

Considering that environmental matters necessarily involve complex legal issues, it is obvious that environmental attorneys play an important role in helping organizations deal with compliance issues; however, many organizations do not fully understand how to use legal counsel effectively or efficiently. Because the objective of sustaining total environmental compliance is to avoid the problems that stem from violations, organizations should emphasize using legal services preventively rather than consulting attorneys only when problems arise. Preventive legal services consist of advising clients on regulatory issues and helping them to avoid exposure to liability. Although many organizations attempt to minimize their legal costs by forgoing preventive services, in the long run the wise use of legal counsel to avoid noncompliance is less expensive than retaining counsel only when enforcement issues become a concern.

The primary premise underlying preventive law is that the timely procurement of legal services will reduce a client's exposure to liabilities as well as future expenditures on legal services. Although many organizations avoid using preventive legal services because they are perceived as being expensive and without tangible return, the consequences associated with environmental enforcement can greatly exceed the cost of preventive services. This is unfortunate, as many persons who fail to use legal services preventively apply the preventive concept in other areas of their lives. For example, most people understand that regularly brushing teeth is over time far less expensive and burdensome than having to endure root canals. The preventive approach to dental care is well understood because the relationship between prevention and the consequences is easily grasped and emphasized through education. Similarly, preventive legal services tend to make sense when managers come to understand how to use them in a way that limits the long-term expenditures.

Preventive legal services in the environmental context can help clients lower their compliance costs by avoiding litigation, preserving remedies, and providing information that facilitates better decisions regarding compliance strategies. For instance, timely assistance by attorneys can sometimes help clients avoid undue regulatory constraints and facilitate favorable determinations by agencies. The key to using legal services efficiently is understanding how and when they can improve a client's ability to operate effectively in regulated environments. One type of service that is relatively inexpensive is using legal counsel to review documents that have compliance implications. For example, review of correspondence that is to be sent to agencies does not take much time and can detect ambiguities and misstatements that can lead to regulatory problems. Other types of documents that warrant legal review are permit applications, contracts, and monitoring reports. Legal counsel should also be consulted when situations arise in which the potential for litigation is evident. Timely advice during the initial stages of matters such as employee problems, contractor and vendor disputes, and contentions by agencies regarding compliance requirements can help clients avoid litigation or reduce its costs. Such advice can also help preserve remedies and strengthen legal positions.

The timing of preventive legal services is critical to maximizing the effectiveness of the advice and minimizing the expense of the service. As a general rule, the costs of legal services increase exponentially as legal matters progress. For example, a legal review of a permit application may cost a few hundred dollars, but preparing comments to address issues in a draft permit that could have been avoided through a better application may cost several thousand dollars. By helping the client address critical issues before they reach the stage where extensive legal services are required, attorneys not only help to expedite regulatory matters but do so at a lower overall cost to the client.

Organizations can also use preventive legal services to gain insights on how they can improve their efforts to comply with environmental laws. Although environmental audits prepared by legal counsel are more expensive than those produced by consultants, they are usually more comprehensive and definitive. Furthermore, the costs of these audits can be reduced by focusing on narrow areas that can reveal where compliance efforts need to be enhanced. One way to do this is to have an attorney review a particular environmental document such as a stormwater pollution prevention plan. These narrow reviews can point out where improvement will probably be needed in similar areas. For example, if an organization has failed to follow the requirements for preparing a stormwater pollution prevention plan, then it also should examine the other plans required by regulations or permits. Similarly, legal counsel can assess whether organizations have implemented compliance programs to the extent needed to be eligible for government programs which waive penalties for self-reported violations of environmental laws.

To gain the full benefit of preventive services, clients should know how to work with counsel to keep the costs down. An important aspect is to maintain a relationship with legal counsel, as familiarity with the organization's operations and culture helps counsel to advise clients more efficiently. Also, legal expenses can be minimized by assuming some of the work involved in procuring services. For example, it is less expensive to prepare draft documents for review and markup than to have legal counsel draft them from scratch. Likewise, it is important to consult with counsel early in a matter in order to obtain advice on how to avoid pitfalls and preserve evidence. Another way to reduce legal costs is to become generally familiar with environmental laws. This not only reduces the time needed for counsel to explain legal ramifications, but it also facilitates advice on how best to deal with pending legal developments. Finally, one of the best ways of keeping legal expenses low is to ask counsel what the organization can do to reduce costs. It is beneficial for both client and attorney to have a common understanding regarding the role of the attorney as advisor and the level of effort expected by the client. By working appropriately with counsel, clients can reduce their environmental costs and improve their ability to achieve total environmental compliance.

11 When Things Go Wrong

Sustaining environmental compliance is not easy, and the complex nature of environmental regulation causes many organizations to experience noncompliance. As has been discussed extensively in this book, the best means of dealing with noncompliance is to avoid it by establishing a foundation for achieving compliance and implementing practices that address environmental requirements. The occurrence of noncompliance not only suggests that an organization's approach to environmental compliance could be improved but also triggers the need for deliberate and expedited action to correct the problem. Organizations that are sufficiently committed to achieving total environmental compliance will address compliance problems promptly and take measures to prevent recurrences. Those organizations that fail to act with the necessary level of diligence will find that noncompliance becomes a routine condition.

11.1 CORRECTIVE ACTION

Government agencies take a dim view of organizations that discover violations and fail to correct them promptly.[8,15] In most cases, sustained noncompliance is grounds for enhancing penalties and in egregious cases may lead to criminal enforcement. Organizations that discover noncompliance should avoid aggravating what may already be a serious situation and begin the process of corrective action immediately. Although prevention is by far the best way to address noncompliance issues, organizations should also develop procedures to ensure that corrective action occurs promptly should a violation be discovered. Otherwise, they risk losing control of the process and expose themselves to more severe consequences.

An important consideration when developing corrective action processes is that there are often practical limitations on how quickly a violation can be addressed. Some violations are simple to correct, but others require considerable time, such as when permit modifications or additional capital equipment are required. Control over the pace of corrective action diminishes further when action by third parties is required. For example, permit modifications are difficult to obtain when the responsible agency is burdened with other projects or is otherwise unwilling to process a request expeditiously. Some violations can never be truly corrected, regardless of the effort organizations may be willing to devote. For instance, regulated parties are often required to fulfill compliance obligations at specific times, such as submitting reports prior to a regulatory deadline or collecting samples during a specified time period. Organizations that fail to meet these requirements cannot go back in time to correct them but can only hope to mitigate the noncompliance by belatedly performing the required activity.

The type of corrective action that will be appropriate depends on the circumstances associated with the violation. Although it is critical to begin the corrective action process promptly, organizations must also deal with pragmatic issues when deciding how to act after a violation is discovered. When a violation can be corrected promptly with minimal cost and effort, the procedures should require that such action be completed immediately. Examples of these kinds of violations are missing labels on containers or equipment failures that can be addressed by simple repairs. Regulatory agencies see little reason why organizations should be troubled by these kinds of noncompliance, as corrective action can be implemented quickly with little direct or collateral expense. Furthermore, they are seldom sympathetic to explanations that a violation was allowed to continue because the persons responsible for correcting it had other duties to perform.[91]

Some violations involve more difficult decisions, such as situations when correcting the noncompliance will require substantial time. In these cases, the corrective action procedures should require the process of corrective action to begin immediately. One of the initial steps should be to define the compliance problem and evaluate potential solutions. For example, a facility faced with overly stringent permit limitations may want to consider requesting a permit modification, slowing down production, installing additional control equipment, or some combination thereof. To expedite such evaluations, corrective action procedures should outline the decision process and identify the parties to be involved. Once a solution is selected, the procedures should describe how the organization will assign responsibility for its implementation and track the progress of the corrective action process.

One of the most difficult situations for many organizations occurs when the only means available to achieve short-term compliance is to shut down operations. Some organizations have attempted to address these situations by establishing policies that require shutdowns whenever necessary to meet environmental requirements. Unfortunately, the pragmatic issues associated with shutdowns make adherence to such policies difficult. Shutdowns that continue for any length of time can threaten the economic viability of a facility and create problems for third parties, such as customers who depend on the goods manufactured at the facility and the employees who depend on wages earned only when the facility is operating. Because the social costs associated with shutdowns can be substantial, even government agencies may favor continued operation over a shutdown when environmental harm is not imminent.

One factor to consider when deciding whether a shutdown is justified is whether continued noncompliance creates conditions that may cause harm to human health or the environment. While organizations should almost always shut down when noncompliance is causing substantial harm, it can be difficult as a practical matter to determine whether actual harm is occurring or is imminent. Because most emission and discharge limits are conservative and do not correlate directly to environmental quality, exceeding them need not be presumed to adversely affect human health or the environment. Furthermore, it can be exceedingly difficult to determine whether a particular release is causing environmental effects. For example, to adequately evaluate the risks presented by an unpermitted wastewater discharge might require the discharger to analyze the pollutants, model the dispersion characteristics of the discharge, and assess the ecological effects. These kinds of practical limitations can make it impossible to estimate actual or potential environmental effects within the time frames in which decisions to curtail operations must be made.

It cannot be emphasized too strongly, however, that the absence of environmental harm does not preclude enforcement for violations of permits and other regulatory limits; therefore, the mere absence of environmental harm does not justify operating in violation, as issues other than environmental quality must be considered. For example, a malfunctioning fuel meter will not in itself cause environmental effects but will negate the use of fuel consumption as a surrogate indicator of emissions from a boiler. An agency could thus construe this kind of noncompliance as impairing a significant regulatory interest and thus a significant violation. The best approach to deciding whether to shut down operations considers all the relevant issues such as legal requirements, disclosures to regulatory agencies, alternate legal mechanisms such as compliance orders, and whether action can be taken to mitigate the violation.

Another important issue for organizations to consider is specifying which persons have the authority to shut down an operation. Regardless of how broadly an organization disseminates shutdown authority, they need to ensure that someone will be available at all times who can authorize the curtailment of operations. Although it is clear in most organizations that senior officers and plant managers have this authority, many organizations fail to consider whether to vest authority to employees such as production supervisors or environmental managers. The degree to which shutdown authority is vested throughout an organization needs to be determined with due consideration of the corporate culture. However, failing to designate persons with this authority can lead to situations where employees act contrary to management wishes because they were not given sufficient guidance on how to handle situations which may warrant a shutdown.

Considering the sensitivity of agencies to organizations that allow noncompliance to continue after discovery, organizations should give urgent attention to resolving issues that arise from agency inspections. In cases where the organization agrees that noncompliance exists, corrective action should begin immediately and proceed expeditiously because the organization's attitude will be a factor in the size of the penalty assessed. Similarly, organizations should consider how they want to respond when an agency's allegations of noncompliance are disputed. In situations where the burden of satisfying the agency is minimal, the organization may want to alter its practices, as little is gained even should the organization ultimately convince the agency or a court that no violation existed. If the agency requests that a penalty should be paid, then the assessment becomes more difficult, as the payment of even a small penalty can lay the foundation for larger assessments should violations be alleged in the future. In these cases, the organization must decide whether it wishes to contest the violation, the size of the penalty, or both.

11.2 CONTINUOUS IMPROVEMENT

Once noncompliance is discovered, the organization should analyze what went wrong to cause the noncompliance and how things could have been done differently to avoid the problem. Similarly, organizations should evaluate whether they responded properly to the noncompliance. Thoughtful analyses of most instances of noncompliance usually result in insights into how the organization can change its practices to ensure compliance in the future. Common changes that are made following noncompliance include modifying equipment, adjusting operating procedures, and revising how employees are trained.

Organizations should be careful to ensure that changes to compliance practices are sufficient in scope to avoid the recurrence of noncompliance. In particular, organizations should resist the tendency to limit efforts to enhance compliance practices to the particular operations or employees associated with the noncompliance, because in many cases a violation is a manifestation of a greater problem. Employee training is a common area where broad action may be warranted following a violation, because if one employee's training was insufficient to ensure compliance then other employees may also need improved training. Likewise, organizations that operate several facilities should ensure that efforts to improve compliance are not directed solely at the facility where the noncompliance was discovered if similar noncompliance can reasonably be expected at other facilities. For example, if an audit discloses that the stormwater pollution prevention plan at a facility failed to incorporate some requirements, the organization may need to assess whether the plans at other facilities have similar deficiencies.

Thoughtful consideration about what can be done to improve compliance sometimes reveals that actions can be taken within several time frames. Organizations should be careful not to ignore long-term actions that will improve compliance even when more immediate responses are needed to address noncompliance. For example, a permittee may find itself in a situation where it needs to change its analytical practices to address a monitoring requirement in the permit. Although changing the analytical practices will address the immediate concern, the permittee could also seek to revise the monitoring conditions when the permit is renewed. By incorporating such issues into the permitting strategy, the permittee can benefit from the experience by improving its short- and long-term prospects for compliance.

Special consideration is required with respect to compliance improvement when a violation can be attributed to some action or inaction on the part of an employee.

Many entities, including the federal government, seem to assume that some sort of discipline is warranted when employees are associated with noncompliance. In cases where employees have neglected or ignored clear responsibilities, disciplinary action may be in order; however, organizations should carefully evaluate the degree of culpability before deciding to discipline their employees. Not all noncompliance can be attributed to the acts of individuals, and in such cases action other than discipline is more likely to prevent recurrences. For example, employees are often blameless when noncompliance results from unexpected equipment failures or changes in regulatory interpretations. Although there is much that organizations can do to avoid noncompliance in these instances, it is often unfair to attribute specific violations to individual employees.

Another issue to examine is why an employee failed to meet a responsibility. Not all employee failures happen because of negligence or a deliberate disregard of environmental requirements. Sometimes problems are caused because the organization has placed too many demands on employees. In fact, one of the risks of implementing environmental management systems is that organizations can overburden employees with too many tangential or irrelevant tasks. Overloading employees is especially problematic when organizations emphasize procedural approaches to environmental management because the employees are sometimes forced to choose between following corporate procedures and satisfying substantive compliance requirements.

A common cause of noncompliance is employees who do not sufficiently understand the applicable requirements and what must be done to meet them. Additional training is a common response to these situations and is usually appropriate, but organizations should reflect on why the prior training was not effective and make changes when appropriate. For instance, training programs sometimes cover an insufficient scope. In addition, organizations should consider whether employees are sufficiently qualified to cope with their environmental responsibilities. In some cases, the training itself may be adequate but the organization may be unreasonable in how it assigns responsibilities to its employees.

11.3 REPORTING COMPLIANCE PROBLEMS

A significant question once a compliance issue is raised is whether an organization should inform regulatory agencies about the matter. In some cases, regulated parties are required by law to report noncompliance. Examples include permits that require the permittee to notify agencies about noncompliance and regulations that require releases of oil and hazardous substances to be reported. Even when disclosing a violation is not legally required, there are reasons why an organization may want to inform the government. Sometimes corrective action will necessarily involve at least an implicit disclosure of noncompliance, such as when an organization discovers that it has unpermitted discharges and must obtain a permit to address the problem. In addition, organizations are sometimes eligible to have the penalties waived or reduced when they voluntarily disclose violations.[92]

Some of the most difficult decisions regarding disclosure involve issues where serious enforcement consequences are likely but will probably be worse should the government learn about the noncompliance through independent means. For instance, an organization that learns though an audit that a facility has tampered with monitoring equipment will face the prospect of possible corporate and individual criminal liability should the matter be reported to the government. However, should the government discover the problem through means such as an employee complaint, a criminal investigation is almost certain. When faced with these kinds of decisions, organizations benefit from understanding their options and the associated consequences.

Although a noncompliant organization should correct their problems as a matter of principle, they need to be pragmatic when considering whether to disclose violations. Although some people are uncomfortable with acknowledging that pragmatism is important with respect to reporting noncompliance, managers are charged with the duty to act with due care in directing the affairs of the organization, and this requires them to make important decisions with conscious deliberation. Therefore, pretending that practical considerations are unimportant with respect to disclosure decisions is counterproductive, as managers need to fully understand the possible ramifications to perform their duties. Furthermore, the various options will expose organizations to different risks which they need to understand to make wise decisions. Ultimately, these decisions should be governed by factors such as the attitude towards citizenship, the extent of compliance problems, and the effect on collateral matters such as ongoing enforcement proceedings. By consciously considering these factors, organizations can base their decisions on a logical assessment of the ethical and pragmatic concerns associated with disclosing noncompliance.

Although some organizations will factor in the likelihood of getting caught when considering whether a violation should be disclosed, this purely pragmatic consideration has proven to be an unreliable basis for such decisions. The main weakness with considering the prospects of discovery is that it is difficult to foresee all the possible means by which misconduct can be exposed. For example, the government has learned of serious environmental violations through means such as complaints filed by passing joggers, disgruntled employees, and environmental groups.[13,19] Sometimes discovery occurs when agencies observe inconsistencies in permit applications or are alerted to problems while reviewing their files for reasons unrelated to the noncompliance. Simply put, the human imagination often underestimates all the means by which a violation can be discovered and should not be relied on when considering whether to disclose problems.

One of the better pragmatic considerations is whether the government will be interested in learning about the violation. Agencies do not have the resources to prosecute every violation that comes to their attention, and staff members may not be too concerned about violations that have little environmental effect and are promptly corrected. In such cases, there may be little to gain by troubling the agency with the information. By promptly correcting minor instances of noncompliance upon discovery, organizations may face minimal risk of enforcement, even if the agency subsequently learns about the matter. Unfortunately, it is not always easy to determine which infractions agencies consider to be minor. One reason for this is that opinions about what constitutes a serious violation vary widely. Issues that may be perceived by the organization as trivial may be perceived by agency staff as quite serious. For example, agencies have sought large penalties for matters that the violating facilities considered insignificant, such as incomplete warning labels, short delays in reporting releases, and poorly organized records. In addition, agencies sometimes face issues not contemplated by regulated entities, such as political pressures to regulate certain areas more intensely.

Organizations also need to consider how failing to report a violation may affect the activities of regulatory agencies. For example, errors in analyzing wastewater samples not only raise the immediate issue of whether the facility has complied with its wastewater discharge permit but can affect agency activities in ways that may not be readily foreseen. Should the agency later use the monitoring data to evaluate the need for water-quality-based effluent limits to determine waste load allocations that affect other dischargers, the discovery that it relied on bad data can severely disrupt its efforts to regulate water quality. While agencies may be inclined to treat intermittent problems such as analytical mishaps leniently if promptly disclosed, they are far less likely to forgive transgressions that impair important agency functions.

In some cases, the failure to disclose violations can set the framework for more serious compliance problems later. For instance, violation of a reporting requirement is often a more serious violation than the underlying violations. Similarly, many permits and regulations require owners or operators of regulated activities to certify that they are complying with the applicable environmental requirements. In cases where the noncompliance has not been adequately corrected, the organization cannot truthfully certify that the facility is in compliance. In such cases, the choices are limited to disclosing the noncompliance or providing a false certification. False certifications can form the basis for serious enforcement actions and should be avoided. Moreover, reporting the noncompliance during the course of a certification basically constitutes a forestalled disclosure where little or nothing is gained by the delay. By considering whether failing to disclose a problem will result in more noncompliance in the future, organizations can avoid aggravating an already difficult problem.

When the decision is made to report noncompliance, organizations should consider how best to communicate the information to the appropriate regulatory agency. The initial step is to determine to which agency the matter should be reported. Although it is common for agencies to share jurisdiction over regulatory matters, the noncompliance should generally be reported to the agency with the primary responsibility to administer the regulatory program. Once the appropriate agency has been determined, the organization needs to figure out to whom in the agency the noncompliance should be reported. It is not necessary or even wise to report noncompliance problems to the section of the agency responsible for enforcement. In many cases, permits and regulations will identify the person to whom reports should be made. In other instances, it is best to direct reports to the section within the agency that is responsible for regulating the activity associated with the noncompliance. Reporting in this way encourages the agency to act on the information in accordance with its normal enforcement evaluation procedures, whereas reporting directly to the enforcement section may increase the possibility of enforcement.

Another important consideration is how best to express the information regarding the violation. There is no single best way to communicate information about noncompliance, and such reporting should be tailored to the particular circumstances. It goes without saying that the information should be reported truthfully and candidly, as false statements are grounds for serious liability; however, the appropriate degree of contrition will depend on the circumstances. When the misconduct is clearly serious, it may be in the interest of the organization to acknowledge the wrongful nature of the act and express remorse to show that the organization is not disavowing its responsibility. In cases where enforcement may not be inevitable, a more reserved and objective presentation of the facts and corrective action strategy may be appropriate. In any case, the tone of the communication should be sensitive to the agency's attitude toward the violation. It is foolish to discount the significance of matters that agencies perceive to be serious. Similarly, there is nothing to be gained by adopting a tone that invites sanctions when an agency feels that none are warranted.

11.4 RESOLVING COMPLIANCE PROBLEMS

Having discovered a compliance problem, organizations must often work with regulatory agencies to effect a resolution. For example, the most feasible means of addressing issues involving unpermitted releases is to obtain a permit, which in turn requires the entity to work with an agency. In most cases where a compliance problem will require time to address, it will be difficult or even impossible to completely resolve the regulatory issues before achieving full compliance. Nonetheless, an organization can usually obtain interim but partial relief that often will meet its needs, but the various forms of regulatory relief have their benefits and drawbacks, and the degree of protection from subsequent enforcement can depend on the mechanisms used to resolved compliance issues.

The least burdensome way to address a compliance problem is to seek informal guidance from regulatory agencies on how to address the situation. Although informal guidance can be helpful and is in some instances essential, it generally does little to prevent the agency or third parties from pursuing enforcement actions later. Even when an agency employee dismisses compliance concerns because they appear inconsequential, neither the agency nor other parties will be bound by such assurances. In other words, absent adherence to the formal procedures for amending regulations or modifying permits, a government agency generally does not have the authority to grant variances to regulated parties. Informal assurances merely evince a nonbinding intent not to take enforcement action.[34]

More formal enforcement mechanisms, such as compliance orders, provide limited protection from enforcement actions but are warranted when an organization needs some protection from enforcement or when demanded by the agency. These orders typically formalize an understanding that the agency will not commence enforcement provided the facility is operated within the limits described in the order. However, whether compliance orders will protect an organization from lawsuits filed by third parties will depend on the legal and factual circumstances surrounding the order and the violations. In many cases, entering into a compliance order will not prevent third-party enforcement actions.[32]

Case Study: ***Citizens for a Better Environment-California v. Union Oil Company of California*, 861 F. Supp. 869 (N.D. Cal. 1994), affirmed at 83 F.3d 1111 (9th Cir. 1996)**

Facts

Union Oil Company of California (UNOCAL) operates a petroleum refinery near San Francisco, CA. In 1991, the regional agency that administers the National Pollutant Discharge Elimination System (NPDES) program issued a wastewater discharge permit that limited the amount of selenium the refinery could discharge after 1993. UNOCAL unsuccessfully appealed the permit to a state agency and thereafter filed an action in state court arguing that the selenium limit was invalid because the agency had failed to follow procedures required by the Clean Water Act. Attempts to resolve the issue achieved fruition in November 1993 when UNOCAL and the regional agency reached a settlement in which UNOCAL dismissed the lawsuit and paid $780,000 to the agency. The agency, in return, issued an order to UNOCAL that extended the deadline for meeting the selenium limit until 1998. Because UNOCAL wanted to avoid a formal admission of liability, the parties decided that the settlement would not characterize the payment as a penalty, although they freely admitted that for all practical purposes the payment did constitute a penalty.

Citizens for a Better Environment-California (CBE) filed a citizen suit in March 1994, asserting that UNOCAL was violating the Clean Water Act by discharging selenium in levels that exceeded the limits in the 1991 permit. UNOCAL argued that the suit was barred because citizen suits are barred when a government enforcement action results in the payment of a penalty and that its compliance with the order issued in 1993 constituted compliance with the Clean Water Act. The trial court refused to dismiss the case for these reasons, and UNOCAL filed an appeal.

Discussion

Both the trial and appellate courts agreed with UNOCAL that the Clean Water Act precludes citizen suits when a federal or state agency is diligently prosecuting a civil or criminal action regarding the same violations. They also acknowledged that citizen suits for penalties are also precluded for violations that were the subject of an action comparable to the administrative penalty provisions of the Clean Water Act. However, they ruled that an administrative action pursued by a state is comparable to the federal provisions only if they provide for public notice, hearings, public participation, and penalty assessment factors. Although UNOCAL had negotiated a settlement that required it to pay $780,000 to the regional agency, the settlement did not preclude a citizen suit because the legal authority relied on by the regional agency was not comparable to the administrative penalty provisions of the Clean Water Act. In particular, the California statute on which the regional agency relied did not authorize the imposition of penalties and therefore lacked penalty assessment factors. Because the regional agency did not enforce the matter in a way comparable to that required by the Clean Water Act, the payment by UNOCAL did not constitute a penalty under the Clean Water Act. In addition, the compliance order did not formally modify the discharge permit, as the procedures for modifying permits were not followed. According to the Ninth Circuit, the order merely evinced the regional agency's intent not to seek penalties for violations of the limit that occurred prior to the 1998 deadline. Because the permit limits were not modified, the environmental group was free to enforce them in a citizen suit.

Because the resolution of compliance issues sometimes requires the payment of penalties, understanding the factors the government considers important when assessing penalties can help organizations formulate strategies to resolve compliance issues. As a starting point, government agencies will assess whether the violator benefited economically by delaying or avoiding the expenditures needed to achieve compliance. If so, then the government will estimate the magnitude of the economic benefit and ensure that this amount is fully recovered by the penalty. In addition to recovering the economic benefit, the penalty assessment will consider the culpability of the violator and the seriousness of the offense. Some of these factors will focus on the social consequences of the violation, such as whether the violation actually harmed the environment. Another important factor is the importance of the requirement to the regulatory scheme. Certain activities such as monitoring and reporting are essential to the government's basic regulatory functions, and violations that compromise such functions are likely to be treated severely regardless of whether environmental harm occurred.

Other factors that influence penalty assessments reflect on the responsibility of the violator. A critical consideration is the degree of willfulness or negligence associated with the violation. Violations that were committed with deliberate forethought are treated more severely than those that occurred because of negligence. Factors that the government will consider when evaluating whether the violation was willful or negligent are the degree of control the violator had over the violation and the ability to foresee the events constituting the violation. The government will also consider whether the violator had actual knowledge of the legal requirement that was violated and any history of prior noncompliance.

Another consideration that can affect a penalty assessment is whether the violator cooperated in remedying the violation. In general, environmental agencies look to how quickly the noncompliance was reported and corrected, as well as the extent to which the violator responded to requests for more information. Organizations that take responsibility for violations and correct them expeditiously tend to be viewed more positively by the government than those who attempt to conceal violations or engage in unjustified litigation tactics.

Another important issue to consider when resolving noncompliance is injunctive relief. The payment of a penalty is rarely sufficient to satisfy agencies and courts if the violator has not achieved compliance by the time the matter is resolved. In such cases, the settlement agreement or court order will require the violator to perform specific actions to achieve compliance. For example, a court may order a party to install pollution control equipment by a certain date. Injunctive relief often entails larger monetary expenditures than the associated penalties.

In some cases, the government may reduce a penalty if the violator agrees to conduct a supplemental environmental project. These projects consist of activities or projects that are likely to improve environmental quality but which the violator is not legally obligated to perform. By entering into a binding agreement to perform such activities, the violator can avoid or reduce the payment of a monetary penalty. There are some restrictions on such projects, and the government typically requires that the activities bear some relationship to the objectives of the environmental statutes that are being enforced. Common supplemental environmental projects include pollution prevention and environmental restoration activities.

11.5 REBUILDING AFTER ENFORCEMENT

Some organizations come to be perceived as chronic violators and become targets of multiple enforcement actions. Although intense and sustained enforcement causes many organizations to reevaluate their attitudes, it can also impede subsequent efforts to achieve total environmental compliance. For example, requirements in settlement agreements to conduct compliance audits or establish whistleblower programs can reveal widespread and even serious compliance issues that have not been addressed in prior enforcement actions. Also, pressure from the government to discipline employees may result in terminations or demotions that reduce the institutional knowledge to levels that impede expeditious change.

Organizations that have faced significant compliance problems should look first at the effectiveness of their existing systems for ensuring compliance. Sometimes the problem may be attitude, but other factors such as inadequate allocations of resources or insufficient talent are common. Once the deficiencies have been determined, the organization should formulate a plan to address them expeditiously. The obvious reason for expeditious action is to avoid new instances of noncompliance and to remedy existing problems; however, expeditious action also portrays a responsible attitude which may be a valuable asset should other serious noncompliance be discovered and reported to the government. Special attention should be given to making employees aware that noncompliance has serious consequences, emphasizing the importance of incorporating compliance into routine activities, and mandating the expeditious correction of noncompliance.

As noted above, the discovery of additional compliance issues can present a serious problem for organizations that have established a reputation for noncompliance. Passivity in the face of probable but unassessed noncompliance is a common but risky approach, as untimely discovery or unanticipated disclosures can disrupt the allocation of compliance resources and increase the possibility of more enforcement. If an organization has a bad compliance history, it may want to acknowledge that other noncompliance is likely to exist and decide how to deal with the issue. In many cases, organizations can gain more control over the process of discovering and correcting compliance problems by auditing pursuant to the attorney-client privilege. This approach will enable organizations to keep sensitive information confidential while developing appropriate plans to disclose and correct violations.

Another issue faced by organizations is how to handle employees who may have had substantive roles in the former noncompliance. Although some people favor the approach of cleaning house when it comes to culpable employees, some degree of caution is warranted, as these employees may be important sources of institutional knowledge which can facilitate corrective action. Furthermore, the enforcement experience often makes these employees far more sensitive to compliance issues than employees who have never suffered through the process. Wholesale replacement of employees can affect internal relationships within organizations if the newly hired employees act disrespectfully toward the established employees. For example, new employees under these circumstances are sometimes unduly critical of the former compliance efforts and express disdain for the accomplishments of prior employees. This can lead to resentment and impair efforts to achieve compliance. Ironically, the attitude of superiority shown by some new employees can cause them to be far less sensitive to compliance issues than the employees they replace.

Nonetheless, employees who find themselves unable to commit to environmental compliance should be terminated or reassigned to less sensitive positions. When the circumstances indicate that an employee is incorrigible or incompetent with regard to compliance matters, strong action is indicated. However, as noted above, a massive shake-up involving several key employees may not be in the best interest of the organization. Instead, organizations should explore whether constructive approaches will be more effective than punitive efforts.

Finally, organizations that have had serious compliance problems should stop trying to guess at which compliance issues are serious and presume they are all important. It can be difficult under the best of circumstances to predict what kinds of noncompliance may be tolerated by agencies and environmental groups, and having a poor reputation makes such assessments even harder. Instead, organizations should go beyond substantial compliance and work at achieving total environmental compliance. By implementing the proper framework and following sound practices, the goal of complete compliance becomes more reasonable and achievable. Achieving total environmental compliance is not easy, but the benefits of regaining control over an organization's destiny will make the effort worthwhile.

References

1. 40 C.F.R. Part 32.
2. *United Paperworkers International Union v. International Paper Company*, 985 F.2d 1191 (2d Cir. 1993).
3. 17 C.F.R. § 229.103.
4. *In re AES Corporation Securities Litigation*, 825 F. Supp. 578 (S.D.N.Y. 1993).
5. 33 U.S.C. § 1367.
6. 42 U.S.C. § 7622.
7. Selingo, J., Ship scrapper won't restart business here, *Wilmington, N.C., Morning Star*, October 1, 1996.
8. Devaney, E.E., *The Exercise of Investigative Discretion*, Office of Criminal Enforcement, U.S. Environmental Protection Agency, January 12, 1994.
9. 15 U.S.C. § 2615.
10. 42 U.S.C. § 7413.
11. 33 U.S.C. § 1319.
12. Macfarlane, R. and Terry, L., Citizen suits: impacts on permitting and agency enforcement, *Natural Resources & Environment*, 11, 20, Spring 1997.
13. Morenburg, P.W., Environmental fraud by government contractors: a new application of the False Claims Act, *Environmental Affairs*, 22, 623, 1995.
14. California Health and Safety Code §§ 25192 & 25249.7.
15. U.S. Department of Justice, *Factors in Decisions on Criminal Prosecutions for Environmental Violations in the Context of Significant Voluntary Compliance or Disclosure Efforts by the Violator*, July 1, 1991.
16. Eisenberg, M., Corporate criminal liability, in *Lawyers' Deskbook on White-Collar Crime*, The National Legal Center for the Public Interest, Washington, D.C., chapter 1.
17. Hartman, B.M. and De Monaco, C.A., The present use of the responsible corporate officer doctrine in the criminal enforcement of environmental laws, *Environmental Law Reporter, News and Analysis*, 13, 10145, March 1993.
18. Nelson, E.T., Strategic use of environmental laws by labor unions: legitimate labor tactic or environmental extortion, *Virginia Environmental Law Journal*, 13, 469, 1994.
19. U.S. Department of Justice, *Conrail agrees to pay $2.5 million criminal fine for intentionally polluting Charles River at Allston site* [press release], July 24, 1995.
20. U.S. EPA, *Clean Air Act Stationary Source Civil Penalty Policy*, U.S. Environmental Protection Agency, Washington, D.C., June 1, 1994.
21. *United States Sentencing Guidelines*, U.S. Government Printing Office, Washington, D.C., chapter 8.

22. Chambers, J.C. and McGrath, P.C., The newest weapon in criminal enforcement of environmental laws: EPA suspension and debarment, *Journal of Environmental Permitting*, 3, 209, 1994.
23. U.S. Environmental Protection Agency, Final Supplemental Environmental Projects Policy, 63 *Federal Register* 24796, May 5, 1998.
24. *The Mumford Cove Association, Inc. v. Town of Groton*, 640 F. Supp. 392 (D. Conn. 1986).
25. *United States v. Sinskey*, 119 F.3d 712 (8th Cir. 1997).
26. *United States v. Park*, 421 U.S. 658 (1975).
27. *United States v. Hopkins*, 53 F.3d 533 (2d Cir. 1995).
28. *United States Sentencing Guidelines*, U.S. Government Printing Office, Washington, D.C., commentary to § 8.A1.2.
29. *In the Matter of Cox Creek Refining Company*, TSCA Docket No. III-421, Office of Administrative Law Judges, U.S. Environmental Protection Agency, December 27, 1996.
30. *Connecticut Fund for the Environment v. The Upjohn Company*, 660 F. Supp. 1397 (D. Conn. 1997).
31. *United States v. Frezzo Bros.*, 602 F.2d 1123 (3d Cir. 1979).
32. *United States v. Smithfield Foods, Inc.*, 965 F. Supp. 769 (E.D. Va. 1997).
33. *United States v. White*, 766 F. Supp. 873 (E.D. Wash. 1991).
34. *Student Public Interest Research Group of New Jersey v. Fritzsche, Dodge, & Alcott, Inc.*, 579 F. Supp. 1528 (D.N.J. 1984).
35. 40 C.F.R. Part 112.
36. *United States v. Weitzenhoff*, 35 F.3d 1275 (9th Cir. 1993).
37. *Sierra Club v. Union Oil Company of California*, 813 F.2d 1480 (9th Cir. 1987).
38. Comment, Limits on individual accountability for corporate crimes, *Marquette Law Review*, 67, 604, 1984.
39. *United States v. Pennwalt Corporation*, Docket No. CR88-55T, filed in U.S. District Court for the Western District of Washington, 1988.
40. *Simon v. Simmons Foods, Inc.*, 49 F.3d 386 (8th Cir. 1995).
41. Werfel, G.L., Crime prevention and mitigation, in *Lawyers' Deskbook on White-Collar Crime*, The National Legal Center for the Public Interest, Washington, D.C., chapter 14.
42. U.S. Environmental Protection Agency, Policy on Environmental Auditing, 51 *Federal Register* 25004, July 9, 1986.
43. Gergacz, J.W., *Attorney-Corporate Client Privilege*, 2nd ed., Shepard's/McGraw-Hill, New York, 1990.
44. Rawson, E.B., Overfiling and audit privileges strain EPA-state relations, *Natural Resources and Environment*, 13, 483, 1999.
45. International Organization for Standardization (ISO), *Environmental Management Systems — Specification with Guidance for Use*, ISO 14001, American National Standards Institute, New York, 1996.
46. 33 U.S.C. § 1341.
47. U.S. GAO, *Water Pollution: Many Violations Have Not Received Appropriate Enforcement Attention*, GAO/RCED-96-23, U.S. Government Accounting Office, Washington, D.C., 1996.
48. Garrett, T.L., Judicial review after *Chevron*: the courts reassert their role, *Natural Resources & Environment*, 10(2), 59, Fall 1995.

49. Brownell, F.W., Regulation by guidance: a response to EPA, *Natural Resources & Environment*, 10(3), 56, Winter 1996.
50. 42 U.S.C. § 6903(27).
51. 49 *Federal Register* 38097, September 26, 1984.
52. 33 U.S.C. § 1362(14).
53. *Weinberger v. Romero-Barcelo*, 456 U.S. 305 (1982).
54. *United States v. Plaza Health Laboratories, Inc.*, 3 F.3d 643 (2d Cir. 1993).
55. 49 *Federal Register* 49568, December 20, 1984.
56. 55 *Federal Register* 50450, December 6, 1990.
57. 56 *Federal Register* 27334, June 13, 1991.
58. 59 *Federal Register* 6126, February 9, 1994.
59. U.S. EPA, *New Source Review Workshop Manual*, U.S. Environmental Protection Agency, Washington, D.C., 1990.
60. 18 U.S.C. § 1001.
61. 18 U.S.C. § 1341.
62. U.S. EPA, Office of Inspector General, *Emission Factor Development*, U.S. Environmental Protection Agency, Washington, D.C., 1996.
63. 33 U.S.C. § 1319(c)(4).
64. 42 U.S.C. § 7513(c)(2)(A).
65. *United States v. Carr*, 880 F.2d 1550 (2d Cir. 1989).
66. 33 U.S.C. § 1321.
67. 42 U.S.C. § 9603.
68. 42 U.S.C. § 11004.
69. U.S. EPA, Office of Solid Waste and Emergency Response, *Questions and Answers on Release Notification Requirements and Reportable Quantity Adjustments*, EPA/540/R-94/005, National Technical Information Service, U.S. Environmental Protection Agency, Springfield, VA, 1995.
70. 42 U.S.C. § 7465.
71. 54 *Federal Register* 27274, June 28, 1989.
72. U.S. EPA, *Limiting Potential To Emit in New Source Permitting*, U.S. Environmental Protection Agency, Washington, D.C., 1989.
73. 40 C.F.R. § 372.
74. 40 C.F.R. § 52.21.
75. U.S. EPA, *Application Form 2C — Wastewater Discharge Information*, EPA Form No. 3510-2C, 1995, U.S. Environmental Protection Agency, Washington, D.C.
76. 40 C.F.R. § 122.
77. U.S. EPA, Office of Water, *Technical Support Document for Water Quality-Based Toxics Control*, EPA /505/2-90-001, U.S. Environmental Protection Agency, Washington, D.C., 1991.
78. U.S. EPA, Environmental Appeals Board, *The Environmental Appeals Board Practice Manual*, EPA 100-B-94-002, U.S. Environmental Protection Agency, Washington, D.C., 1994.
79. American Public Health Association, American Water Works Association, and Water Environment Federation, *Standard Methods for the Examination of Water and Wastewater*, 18th ed., American Public Health Association, Washington, D.C., 1992.
80. U.S. EPA, Environmental Monitoring Systems Laboratory, *Methods for Chemical Analysis of Water and Wastes* (EPA-600/79-020), U.S. Environmental Protection Agency, Cincinnati, OH, 1983.

81. AOAC, *Official Methods of Analysis of the Association of Official Analytical Chemists*, 15th ed., Association of Official Analytical Chemists International, Gaithersburg, MD, 1990.
82. U.S. Geological Survey, *Methods for Determination of Inorganic Substances in Water and Fluvial Sediments*, U.S. Geologic Survey, U.S. Department of Interior, Denver, CO, 1979.
83. ANSI, *American National Standard for Photographic Processing Effluents*, American National Standards Institute, New York, 1975.
84. 45 *Federal Register* 33314, May 19, 1980.
85. U.S. EPA, *Compilation of Air Pollutant Emission Factors*. Vol. I. *Stationary Point and Areas Sources*, 4th ed., U.S. Environmental Protection Agency, Washington, D.C., 1985.
86. U.S. EPA, Office of Solid Waste and Emergency Response, *Test Methods for Evaluating Solid Waste (SW-846)*, 3rd ed., U.S. Environmental Protection Agency, Washington, D.C., 1986.
87. U.S. EPA, Office of Water, Engineering and Analysis Division, *Guidance on Evaluation, Resolution, and Documentation of Analytical Problems Associated with Compliance Monitoring*, EPA 821-B-93-001, U.S. Environmental Protection Agency, Washington, D.C., 1993.
88. U.S. EPA, Environmental Monitoring Systems Laboratory, *Methods for Chemical Analysis of Water and Wastes*, EPA-600/79-020, U.S. Environmental Protection Agency, Cincinnati, OH, 1979.
89. 56 *Federal Register* 50758, October 8, 1991.
90. U.S. EPA, *Source Assessment: Phosphate Fertilizer Industry*, EPA-600/2-79-019c, U.S. Environmental Protection Agency, Washington, D.C., 1979.
91. *United States v. Starr*, 535 F.2d 512 (9th Cir. 1976).
92. U.S. Environmental Protection Agency, *Policy on Self-policing: Discovery, Disclosure, Correction and Prevention of Violations*, 60 *Federal Register* 66706, December 22, 1995.

Index

A

C

D

E

For Product Safety Concerns and Information please contact our EU representative GPSR@taylorandfrancis.com
Taylor & Francis Verlag GmbH, Kaufingerstraße 24, 80331 München, Germany

www.ingramcontent.com/pod-product-compliance
Ingram Content Group UK Ltd.
Pitfield, Milton Keynes, MK11 3LW, UK
UKHW011011180926
475701UK00004B/684
* 9 7 8 0 3 6 7 5 7 9 0 2 9 *